The Making, the Rise, and the Future of the Speakingman

Fourth Edition

Revised and Updated

Dan M. Mrejeru

I would like to dedicate this book to the memory of **Spiru Haret** (15 February 1851-17 December 1912), who was a Romanian mathematician, physicist, and astronomer. His contribution to celestial mechanics, to which he provided a third-degree approximation for the distribution of forces in major instability of major axis of orbits, was a world first achievement.

In his work he provided a result fundamental for the n-body problem, and he was **a world pioneer of nonlinear dynamics**.

After his return to Romania (1878), he **dedicated the rest of his life to improve Romanian education** as a professor and politician.

My grandfather, Leon Mrejeru, a teacher by formation, was greatly influenced by Spiru Haret approaches toward education, and he made himself the most prolific school builder of interwar Romania.

I am a nonlinear person to whom nonlinear wisdom greatly inspired me, while sometimes biased a little, as nonlinear does, my life and my work options.

I would equally like to dedicate this book to **my daughter Anamaria**, who holds a postdoc in Neuroscience, and who, because of her scientific career, raised my highest interest for Neuroscience, Biology, Genetics, and the subsequent **evolution of the modern brain**.

Anamaria's work example made me self-train as an independent researcher and science author. I owe her all my research work.

Published in the United States of America

ISBN 979-8-89395-816-4 (SC)
ISBN 979-8-89395-815-7 (Ebook)

LCCN 9781957041773

A Terrestrial Mind Publishing, LLC
1710 W Touhy Ave.,
Unit E Park Ridge IL 60068
www.danmirceamrejeru.com

Order Information and Rights Permission:

Quantity sales. Special discounts might be available on quantity purchases by corporations, associations, and others. For details, contact the publisher at the address above.

For Book Rights Adaptation and other Rights Permission.
Call us at toll-free 1-888-945-8513 or send us an email at admin@stellarliterary.com.

Book's Abstract

This book collects many articles that I uploaded on academia.edu in 2022/23.

This collection enhances **the most significant moments that**, in my opinion, **have occurred in human evolution**.

-a general climate cooling occurred from 7.0 to 5.5 million years ago.

-after the cooling, **changes from subtropical forests into savannas** (Sahara and other deserts were also formed at this time) **made the first hominins appear 5.4 million years ago**.

-other major planetary changes occurred around 1 million years ago:

a) it intervened **a change in the glaciation cycles** (the previous 40,000 years cycles with equal warm/cold eras turned into 100,000 years cycles with 90 percent of each cycle represented by glaciation).

b) **a geomagnetic change** (last polarity long-term change took place 773,000 years ago) where the geomagnetic excursions replaced the long-term reversals, which previously occurred every 22 ka; new pattern of excursions would have or not short-term reversals).

-during the geomagnetic excursions the average geomagnetic strength diminishes from an average of 45,000 nT to 10% of it or around 4,500 nT, and even further if a temporal reversal occurs (3,000 nT); it influences the brain connectivity.

-the **fire-use and tool making were identified within the era of the Last Clade** (tool bifaces persisted until 200 ka, fire use shows frequent hearts only 100 ka; fire use and cooking did not affect the increases in hominins brain volume).

-the hominin brain begins to differentiate from monkey only in the last 450,000 years that is called the Hominins Last Clade.

-among changes in the hominin brain, it develops **the process of self-domestication** that incipiently started 600,000 years ago, but **it took a long time to mature within the societies around 400,000 years ago**; self-organization is an assumable **result of forming cooperative**

societies; even then, the socialization implied the development of a language (a raw, primitive, and static language, being doubled by a sign-language).

-the expansion and rearrangement of temporoparietal junction and posterior superior temporal sulcus (TPJ-PSTS) have played **a key role in the growth of human social abilities**; this process is assumed to have started developing some 450,000 years ago.

-Neanderthals left little in the range of symbolic art, and this is an indication that the **Neanderthals have been capable of some speech that produced an incomplete range of sounds**.

-ROS (Reactive Oxygen Species) level progressively increase in the neurons during cortical development; ionizing radiation increases the cell surface expressions and instantaneously generates ROS as hydroxyl radicals; such radicals are rapidly de-excited (deactivated) by other protective processes, and de-excitement produces biophotons; embryonic neural cells possess relatively high cellular ROS levels that promote their proliferation and survival.

-large mammals extinctions tend to occur when the geomagnetic strength is weak (like during the excursions and short-live reversals).

-it is postulated that reversals generate ozone holes, which can be penetrated by harmful UV (Ultraviolet) radiation, causing extinctions.

-it is also postulated that the **faunal change from large mammals to smaller ones was caused by a change in the natural strategy of hunting among species**.

-there is strong scientific evidence that **ROS pulsations are associated with neurogenesis and have been the determinant factors which generated changes in the hominin brain architecture in the last 450,000 years**.

-fire-use did not require intelligence but noncognitive skills; the same applies to tool making that evolved into advanced tools only 40 ka.

-the **Homo sapiens brain entered the path toward modernity only 70,000 years ago and precuneus appears to be the most significant change in the new anatomy of human brain**; precuneus evolution represents an extension into it of the neighboring brain structures as a novel process of self-organization.

-the parietal lobe has a crucial role in object construction; the posterior parietal cortex (PPC) generates visomotor functions, named visually based imagination, where the eye-hand coordination allows fine control of hand force manipulation and allows mentally driven spatial constructs; the medial parietal cortex as a whole is essential in high-order cognitive processes, autobiographic memory retrieval (the self), the coordination of processes causing social perception; all these new functions appeared only in the last 30-23 ka.

-Precuneus evolves during the entire lifetime and differs between individuals and races.

-it is demonstrated that the swarming of species is generated by the van der Waals weak forces.

-the role played by "latency" in functioning of the human brain is underestimated or largely diminished: it implies a reduction in the brain energy consumption and a lower entropy.

-the **hierarchical change of our mental complexity and its implications are mostly omitted**.

-the **high dimensional origin of emotions** and their translation and alteration in the conscious state is very little explained.

-changes that generated the modern brain have been directed to avoid extinction because they brought to the brain a novel self-organization.

The book insists on the epoch 110,000-23,000 years ago, when the most anatomical changes occurred in the Homo sapiens brain. Other biological evolutions are considered.

Part two makes several estimates on the determinants of human migration: the migration out of Africa may have been facilitated by four distinct trends, which may have been used alternatively or simultaneously.

Part three debates evolutionary biological events, which shaped the modern brain.

Part four analysis the evolution of conscious, unconscious, human thinking, and attempts to map consciousness.

Part five revises some novel ideas on neurogenesis in association with geomagnetic events and solar cycles.

Part six debates the most significant mental features, which may affect our future.

A conclave on Homo loquens evolution

This is intended as a revised and updated **abstract** of the book

1. The book identifies two distinct cognitive revolutions:
   -the first revolution that evolved toward a nonlinear thinking
   -the second revolution that generated current linear thinking
   It seems that the tubular tracts for the transmission of perceptual signal information appeared relatively recently in the Homo sapiens section between Temporal lobe and Occipital area. In monkeys, this particular section is served only by multiple synapses. There is experimental evidence that synapse-to-synapse transmission of signals generates nonlinear processing and outcomes. Also, experimental evidence suggests that the longer the range of transmission in the tubular tract is, it induces more linearization of processing because of a longer section of friction with the tract-wall causes more top-down processes and slows down the signal transmission speed.
   Top-down perception refers to situations in which context, learning, or expectation alters a perceptual process. It works downward from initial impressions down to particular details.
   In ambiguous situations, generated by a nonlinear perception, the brain needs to fill in the gaps.
   Top-down processing takes a higher-level representation as input and produces a modified low-lever representation as output. Because information is organized as hierarchies, the lower level of the hierarchy represents the detailed information. As it seems, V2 encodes only top-down control settings, and V1 encodes only bottom-up salience.

The order of visual processing is a top-down process, with higher order forms processed first, fo llowed by lower-order forms.
In short, the top-down breaks the whole into parts, while the bottom-up combines simple into complex.
The first cognitive revolution was enhancing the bottom-up processes (70,000 years ago), which created mental complexities (as nonlinear outputs) from simple (linear) details, while the second revolution did the reverse, enhancing the top-down processes.
The top-down process must be responsible for producing a rudimentary linearity that assures surviving in all species.
But around 70,000 years ago, evolution enhanced its opposite that is the bottom-up, stimulating a nonlinear ambiguous thinking.
And then, around 30,000 years ago, the top-down processing intervened to eliminate the ambiguity, while causing linear predictability.
The top-down processes seem to serve long-range cortico-cortical connections; the bottom-up processing is supported by short-range circuits.
Here, **one can hypothesize that in the last 100,000-70,000 years, the Homo sapiens and other hominins evolution alternatively adapted to two distinct processes, which are the bottom-up and top-down processing. Presumably, these alternative adaptations were guided by the evolution of short-range and long-range neural synaptic connections**.
In fact, both adaptations represented novel tentative and solutions to reorganize the perceived hierarchical information to fit new environmental constraints.

2. We became a new species (in the epoch 70,000 to 23,000 years ago) that rapidly transformed nonlinear mental approaches (which appeared 70,000 years ago) into an ADVANCED LINEARIZATION (that appeared around 30,000 years ago).
The **prefrontal cortex (PFC) and parietal areas are sections of the brain developed in the last 70,000 years**. Precuneus/posterior cingulate cortex are key components of a

network that involves an extension of spatial navigation to help with construction of scenes or situation models. A situation model allows the viewer to construct a rich percept of a place. **It generates internally maneuvered cognitive scenes or contexts and relates them to an external context**.

Hence, **only 70,000 years ago this capacity to generate mental spaces begun to form, and it later generated a two-dimensional (planar) mental space that has "linear" properties (30,000 years ago).**

However, this "linear" two-dimensional thinking remained solely expressed until 1830 as a Euclidian Geometry, or a Planar Geometry.

In 1829-1830 Lobachevsky and Bolyai introduced the concept of hyperbolic geometry (as a three-dimensional geometry).

Archaeologically, it was identified <u>a bulging and rounding of the parietal lobe only 37,000-35,000 years ago</u>.

It is assumed that the mentioned bulging is related to a significant neurogenesis that occurred in the striatum.

Dr. A. Vyshedski insists that **a language-ready-brain implied a full development of PFC (Prefrontal Cortex), and it did not appear before 35,000-30,000 years ago**.

The development of precuneus favored the evolution of dynamic approaches in Homo sapiens brain, and this new approach was able to generate the novelty of verbs and adverbs, essentially changing the expression and utility of the language. For the first time it appeared the concept of the flow of time (as present, past, and future).

Also, **the intraparietal sulcus (as a further development of parietal lobes), and supramarginal gyri in the inferior parietal lobes, and areas of prefrontal cortex, as a neurological substrate, define "quantity appreciation, number processing and topological representation of numbers", while it allowed abstractions**. Numerology seems to be an inherently abstractive process. It has been speculated **that lower-level abstraction involved in numerology could serve as a fundament for higher level symbolic thinking**.

This is why abstract thinking, as the initial quantification approach (as it introduced decontextualization and low-level numerology), appeared before the manifestation of symbolic expression (as high-level numerology).

This was **the first separation between "things"** (to which numerology was applied) **and feelings, emotions, or sensations**.

Assumably, the mentioned **separation occurred sometimes around 45,000 years ago**, as it becomes expressed in the cave paintings of the South Sulawesi (Indonesia), but only 10,000 years later in Western Europe, and nowhere else in the rest of the world in this timeframe.

There is very hard to define a separation line between monkeys and humans. This is so because the monkeys can express abstract thinking, and in part, they could represent small numbers symbolically. However, humans show the capacity to represent explicit high-level symbols, like large numbers. Hence, we can symbolically manipulate such large numbers, producing mathematics.

Firstly, the human abstraction capacity is initially recorded 70,000 years ago in Blombos Cave of South Africa. It failed to evolve further until it appeared again 45,000 years ago in a cave in Soth Sulawesi. It took another 10,000 years to appear in Western Europe.

As the cave painting analysis indicates, the **symbolic art appeared somewhere between 30,000 and 24,000 years ago**. All previous cave painting expressed only abstract thinking.

Abstract thinking existed in the human mind for almost 45,000 years, but it manifested only in isolated outcomes, which were separated by large intervals of time.

Much better represented was the industry of figurines and other similar artifacts.

As the cave archaeologists indicate, the abstract manifestation ceased in cave painting around 25,000 years ago, being replaced by symbolic thinking.

Therefore, Homo loquens anatomy was constructed from 70,000 to 30,000 years ago, but it maturated and became fully expressed only 25,000 years ago.

Most neuronal circuits in the human brain are hierarchically organized which allows various brain sections to share some features.

Here, there is very interesting to find that some newly developed areas of the brain, like prefrontal expansion, parietal expansion, cerebellar areas expansion, and especially precuneus unique evolution, turned to reach a top hierarchical place in brain organization, and exercise a commanding or dominant role in novelty generated functions of the brain architecture.

However, another separation line was drawn by dynamic language acquisition some 30,000 years ago.

Assumably, these two dates (30,000 to 25,000 years ago) may amount for the divide (the theoretically assumed divide is clearly manifested 27,000-26,000 years ago).

This implies that mastering of the dynamic language took many thousands of years.

The mentioned divide irreversibly separates Homo sapiens from Homo loquens.

Many features of modernity are expressed only in Homo loquens, while some little others appeared in hybrids.

The **abstract decontextualized human thinking had evolved, but significantly much later, into logic thinking (during Greek Antiquity)**.

Even then, contextual thinking remains today valued and preserved in Eastern Culture.

However, the **symbolic cave art was not a PFC manifestation**, but a transitory one. This defines the (coherent) dynamic language (as opposed to previous static language that had no verbs) as a mental outcome no older than 30,000 years or 24,000 years.

Therefore, **between 70,000 to 23,000 years ago the pre-frontal and frontoparietal developments, unique to humans, represent the origin of "linearization" and "quantification." It also is the transition interval from Homo sapiens to the new species I named Homo loquens (the speakingman)**.

I can say that **Homo sapiens migrated 55,000 years out-of-Africa as a hybrid, and during its Eurasian expansion, this**

hybrid gradually transformed into the new species of Homo loquens.

A solid expression of Homo loquens manifested some 25,000 years ago.

By several back-to-Africa reverse-migrations, it was completed the **African hybrid Homo sapiens evolution to a full Homo loquens new species**.

However, Homo sapiens, as a hybrid species, entered New Guinea and Australia approximatively 48,000 to 50,000 years ago. Those Homo sapiens were hybrids. Also, the early Middle Easterners and Europeans were hybrids, too.

All these hybrids did not speak a dynamic language. It took another 15,000 years for a dynamic but linear language to appear among the representatives of the new species of Homo loquens.

George Poulos, a linguist at the University of South Africa said that "*the first speech sounds came a mere 70,000 years ago, and the ability to produce vowel and consonant sounds didn't evolve until around 50,000 years ago*."

New findings agree with recent genetic studies that show changes in Homo sapiens, since its split from Neanderthals, and they demonstrate that Homo sapiens is an evolving species with deep African roots.

3. When all evidences are evaluated, it results that **"linearization" is the main aspect that generated all other elements of "modernity" and created the new species of Homo loquens**.
4. According to Wikipedia "Acoustics is absolutely rife with linearity. Nearly every acoustic system is linear to a high degree. For most sounds you hear, air is a linear medium.
5. In a study of Thomas Deneux, Evan R. Harrell, Alexander Kempf, Sebastian Ceballo, Anton Filipchuk, and Brice Bathelier, titled *Context-dependent signaling of coincident auditory and visual events in primary visual cortex*, published on May 23, 2029, in eLife (doi;10.7554/eLife.44006), the authors indicate that experimentally was demonstrated that sound inhibits the visual processing of visual information, altering the context. It

demonstrates how dynamic language processing inhibits the processing of the information content present in contexts.

In sum, these data show how "linearization" had developed in the human brain in the last 30,000 years and its direct connection with the development of dynamic language.

Another study from Hebrew University, Jerusalem, Israel, published on Feb. 13, 2008 (doi:10.1152/jn.00469.2007) has demonstrated that "modulating the sound amplitude in counter phase to the two years generally produces weaker modulation of the optical signal than the modulation produced by the monaural stimuli. Hence, sound used by language tends to diminish the intensity of the optical signal (that is nonlinear), favoring "linearization."

One of such dominant new features is the transformation of qualities into quantities. Here, language is subdued to the same type of transformation.
Since 45,000 years ago, humans manifested an incipient tendency to **transform qualities into quantities** by drawing raw geometric forms of animals as symbols. Planar geometry appeared as an approximation of natural shapes, and as an attempt to quantify these natural shapes.

6. When the number of neurons has increased in the human brain, such an increase is directly correlated with an enhancement of cognitive abilities. However, the evolution of human cognition has been accompanied by a process of prolongation of neural processing because of an occurring **accumulation of processing time of each individual neuron that is called "latency."**

The process of stretch growth is an axon growth mechanism. As animals grow, it takes place a rapid expansion of the nervous system directed by mechanical forces on axon tracts.

Forces implied in growth stimulate the axons to add cytoskeleton, axolemma, and other cellular building materials along the central length of the tracts to minimize strain. Otherwise, the axons would be stretched to the point of rupture.

In addition to creating axons of exceptional length, this process can exceed by far the well-established processes of axon sprouting and regeneration.

Potentially, the process of elongation helps to rapidly transport in the axon's cylinder and fibers the building blocks of the brain and redistribute them in required brain sections.

The tracts evolve through the lifetime and differ from one person to another. Similarly, precuneus evolve through lifetime and from one person to another.

For example, when cortical neuronal processing increased in auditory system, **it became evident a benefit vs. disadvantages of slow cognitive reaction. It was discovered that a longer time window for cortical processing is advantageous for analyzing time-varying acoustic stimuli, such as those important for speech processing. In short, the latency exposed much more variation, aka details of the sound**.

Our modified brain architecture produces "functional latencies", which result from neuronal multiplication that caused the elongation of neuronal tracts; the elongation diminishes the tract diameters, increases the tract's wall resistance, while slowing down the speed of the perceived neural signals; it makes us see the previously unseen that was embedded within an unaltered speed.

However, some of the studies indicated that rather than thinning down due to stretching, the diameter of axon remains approximately the same and the cytoskeleton appears normal.

The "latency" also may contribute to diminishing tract's entropy (energy), where the increased friction, producing energy, implies the excess energy dissipation into the neuronal environment. This made the "primary" consciousness of the Late Paleolithic era show higher overall entropy. Today, neuroscience research associate's higher entropy with intelligence.

The monkeys show functional "latencies" of 50 milliseconds to 100 milliseconds; a human speaker, during a conversation, may show functional latencies in the range of 1,500ms to 3,500ms.

The effects implied by our new species evolution.

Linearization:

-it generated the linear language around 30,000 years ago; the language became recurrent, when sound processing became developed based on recurrent properties, allowing assemblage of mental compositions.

-rapid evolution of novel linear approach vastly extended the consciousness that is exclusively limited to linear processing.

-it produced the linearization of emotions ,which led to self-domestication as the result of altering the raw emotions; self-domestication increased the capacity of socialization of our species.

-it produces switching the qualitative aspect of the emotions toward a quantitative aspect.

-it gradually developed the mental analysis and synthesis, leading much later, or during Greek Antiquity, to logic thinking.

-it gradually diminished the role of collectivity and context within the Western Culture, indicating different rates of evolved "linearization" between the Western and the Eastern Cultures.

Gradual transformation of the perceived nonlinear qualities into linear quantities, which can be analyzed and mathematically processed (statistically) at will.

-original simple quantification as associated with initial form of numbering

-analog quantification

-digital quantification

-Artificial Intelligence that attempts to quantify the thinking system.

Functional "latencies."

Various studies found that the number of neurons in humans was only 10 percent increased until 10,000 years ago compared with other mammalians. One study found that that mentioned number of neurons turned to 10 percent decreasing in the last 3,000 years.

In my opinion, beyond the body growth, **the increased number of neurons have produced bigger "latencies." Current assumed reduction in the number of neurons brings us on par with the rest of mammalians, but also reduces the length of previous "latencies."**

However, this opinion may contradict, or it could not explain the evidence that, in the last 200 years, the path of "discoveries" increased exponentially that implies a significant increasing in functional "latencies."

In the meantime, I would like to mention that a group of physicists and neuroscientists from the University of Frankfurt (Germany) have proposed that, in order to reduce the brain energetic cost, the **neurons recently reduced their processing speed**. In this case, the **overall functional "latency" increases disregarding the overall number of neurons or the growth factors**. Also, such a reduction in the processing speed may help explain that 10 percent reduction in the number of neurons.

Maybe the **reduction in the neuron processing speed could have increased in the last 3,000 years as new research indicates and this contradicts the previous finding that refers to** the last 10,000 years. And **maybe this reduction became significantly accelerated in the last 200 years**.

"Latency" characteristics:

-diminishing the speed of neural signal transmission in the neural tracts (produced by the tracts elongation caused by neuronal multiplication) that reduces the speed of neural processing.

-it leads to the **disclosure of those details**, which cannot be analyzed at higher, while normal speeds; at higher, normal speeds, the brain analyzing must reduce the amount of information to the size it can process, and this results in a type of information compression, where many informational details are compacted into larger ones; this compaction causes **what we call the UNSEEN.**

-it leads to produce DISCOVERIES, which are those details embedded in the unseen; as the "latencies" increased, especially as the language evolved, the number of "discoveries" increases exponentially.

-our “**progress” is linear** and results in mounting of novel details into larger complexities, which cannot be resolved by linear thinking.

-our “**progress**” maneuvers the unseen details, which are the past occurrences; hence, it **evolves “backwards” that is not toward the future but toward the past**.

-the future is embedded only in the EMERGENT BEHAVIOR OF EACH EVOLVING COMPLEXITY, BUT OUR MIND CANNOT UNVEIL THE EMERGENT BEFORE OCCURRING IT; thus, we cannot know a priory the future because our prediction is linear and maneuvers only the past occurrences.

A clarification of some of the mentioned aspects.

1. **EMOTIONS are an emergent and complex aspect** generated by the experience accumulated over many generations. It is hard to equivalate the long-term accumulation with regular memories, which are essentially “linear.” Hence, the “**emotions” are stored as nonlinear encounters, but when they are processed by the consciousness, they become “linearized”**.

1. **The emotions cannot be reduced linearly, but even then, our consciousness does such a reduction that significantly alters the content of the complexity embedded in emotion.**
2. **Emotion is therefore nonlinear and complex, and the result of this evolving complexity that arises in the “emergent behavior.”**
3. **Paleolithic mind was 100 percent emotional**, where raw emotions (probably still nonlinear in nature) dominated the mind wholeness, and processed the **emergent behavior** enclosed into them, which **became expressed as CONTEXT**.
4. **Modern brain**, especially in the Western Culture, alters original context of the emotions, while it **ignores this CONTEXT** by producing a linear analysis/synthesis that we call LOGICS, or “logical thinking.”
5. **MODERN BRAIN THINKING REMAINS, AT LEAST 95 PER CENT, THE SUBJECT OF PROCESSING EMOTIONS**. Linearizing it, we introduced linear emotions, which are a linear reduction of original nonlinear, complex emotions stored in the brain.

Conclusion

Analysis of top-down and bottom-up neural processes shows a direct connection to short-range and long-range neural synaptic processes. The bottom-up is dependent on the short-range evolution of neural circuits, and top-down relates to long-range circuits.

In both cases, the bottom-up and top-down deal with hierarchical processing of information. The short-range circuits interact only high-level representation with another high-level representation. By contrast, a long-range circuit can translate a high-level representation into a low-level representation, meaning, it translates a high-level information into a low-level detailed information. It "linearizes" the nonlinear content of information.

During the hominin recent evolution, the last 100,000-70,000 years, it occurred a variety of environmental constraints which forced the hominin's brain to adapt by reorganizing the perceived hierarchical information to fit the imposed constraints.

I hypothesized that around 70,000 years ago the neural reorganization relayed on bottom-up processing. The result was a slide toward nonlinear thinking, triggered by the short-range neural circuitry development. However, the generated ambiguity of meanings was not provided a right fit, and continual environmental changes forced to another tentative, as a correction to previous reorganization. It introduced a novel linear thinking, but which was biologically stimulated by the long-range neural circuits development.

A study by Christophe Bossens and Hans P. Op de Beeck published online by Frontiers in Behavioral Neuroscience found that, in rats, "*nonlinear tasks cannot be trivially solved by simple V-1 model.*" They found that "*rats are able to solve linear features tasks but are unable to acquire the nonlinear features*."

This and other studies conclude that most nonlinear features in various animals are not addressed because a strong inhibition occurs.

As it seems, in the last 100,000 years, the hominin brains made a complex adaptation of processing, reorganizing twice their hierarchical perceived information in order to overcome some existential environmental risks. The last linear reorganization favored the long-range neural connections and helped produce high-order cognitive skills (planning and emotion regulation) by entangling far-reaching networks.

Could we expect another reorganization of hierarchically perceived information that may balance the linear and nonlinear neural processing or to balance predictability with uncertainty?

Contents

A conclave on Homo loquens evolution viii

Part One Making of the Speakingman 1

Making of the Speakingman 2

Part Two The Migration 51

Why did prehistoric people migrate out-of-Africa? 52

Magnetoreception may have contributed to ancient migration. 63

Could the geological fault lines have served as a guide to ancient human migrations? 69

Part Three Rising of the Speakingman 74

Rising of the Speakingman 75

Part Four How did evolve the Speakingman thinking? 89

Information transition from nonlinear of unconscious to linear of conscious and the overall evolution of consciousness 90

The evolution of unconscious processing from monkeys to humans 108

Chaos, nonlinear, and linear in human brain 125

Our civilization originates into a hierarchical regression of vertical complexity 136

Mapping the consciousness 154

The First (70,000 years ago) and Second (30,000 years ago) Cognitive Revolutions 184

Emergent qualities and quantities as information 184

Part Five Some thoughts on Neurogenesis role on Human evolution, as a result of forcing of Solar Cycles and the Geomagnetic Excursions 214

A unique neurogenesis produced the modern brain......................... 215

A hypothesis on Human evolution, as the result of forcing by Solar Cycles and Geomagnetic Excursions 223

Part Six The Future of the Speakingman 240

The Future of the Speakingman 241

The origin of human quantification and its relation to human language 267

Is our future a slide into total individualism? 277

The Role of neural functional latency in the evolution of the modern linear human brain 298

Two different modalities to approach discovery. How our "progress" can run simultaneously backward and forward 314

A reason to catch emergent behavior: it shapes our future. 327

About the Author 337

Part One

Making of the Speakingman

Making of the Speakingman (I dedicate this section to my previous book Solovki's Ersatz)

(An Anthropological review based on the Science of Complexity interpretation and the occurrence of Geomagnetic events)

Abstract

The aim of this paper is to demonstrate that human evolution was highly dependent on two distinct stages of the brain organization.

In the first stage have been created the hominins as result of a major anatomical change in the functions of the brain. This stage occurred in the era 500 ka to 300 ka, and it generated a cerebellar/encephalization revolution that resulted in the first hominins who were critically distinct from apes: Homo heidelberensis and Homo neanderthalgensis. As it seems, it also took place a split from H. neanderthalgensis that produced Homo sapiens (around 200 ka).

The second stage of some major anatomical changes have occurred within Homo sapiens in era 100 ka to 27 ka, and it created the "modern brain."

As this paper postulates both major anatomical cerebral changes were produced by the intermediation of geomagnetic events, which increased the atmospheric concentration of C14 isotopes, stimulating Reactive Oxidative Species (ROS) and associated pulses of neurogenesis. Hence, neurogenesis was solely responsible for producing encephalization, new cerebellar and frontoparietal structures and significant circuitry reorganizations in the brain.

The paper analyzes various aspects of evolution as the climate and geomagnetic forcing factors to other functional evolutionary aspects, like bipedalism, the use of fire, and Anthropogenic reorganization of the environments.

Bipedalism

On September 30, 2016, journal ScienceDaily published an article generated from the materials provided by Brown University; the article title is *Ancient global cooling gave rise to modern ecosystems*.

The Brown University study, led by Timothy Herbert, a Geology professor at the Department of Earth, Environmental and Planetary Sciences, was initially published in *Nature Geoscience*.

The study explains that: A*round 7 million years ago, landscapes and ecosystems across the world began changing dramatically. Subtropical regions dried out and Sahara Desert formed in Africa. Rain forests receded and were replaced by the vast savannas and grasslands that persist today in North and South America, Africa, and Asia*."

This phenomenon was "*driven by a sharp reduction in atmospheric carbon dioxide*." "*The record revels (based on global sea surface temperatures spanning the past 12 million years) a distinct period of cooler sea surface temperatures spanning 7 million to 5.4 million years ago, the end of Miocene epoch. The global climate during Miocene is known to have been much warmer than the present. During the cool period detected in the study, sea surface temperatures dropped to near modern levels. We have proven that it was a decline in CO2. It could be that there were large-scale geological changes occurring at that time that affected the carbon cycle*."

How such dramatic change in African climate affected the first hominins inhabiting this continent?

Here, I would like to mention a ***Hominid Species Timeline*** published online by Washington University. The sources of information have been Arizona State University and the Smithsonian Institution's Human Origins Program.

According to this timeline the first hominin species was ***Sahelanthropus tchadensis*** that lived 7 million years ago. This species shows the sign of bipedalism. But this species incipience corresponds with the mentioned dramatic cooling in Africa that transformed many forested area into savannas. Thus, many African monkeys were forced to embrace a bipedalism approach or perish.

Next hominin species was ***Orrorin tugenensis*** that lived from 6.1-5.8 million years ago. Next one was ***Ardipithecus ramidus*** from 5.8 to 5.5 million years ago.

Now, from 5.5 to 4.4 million years ago, no other hominin species was found in the archaeological records.

According to the article mentioned at the beginning of this paper, the cooling phenomenon developed from 7 million years ago until 5.4 million years ago. Thus, the first three species of hominins lived in this gradual cooling era.

Next species on record is ***Ardipithecus anamensis*** (4.4-3.8 million years ago), followed by ***Australopithecus afarensis*** (3.8-2.7 million years ago). Next is ***Kenyanthropus platyops*** (3.5-3.2 million years ago), followed by ***Australopithecus africanus*** (3.0-2.4 million years ago) and ***Paranthropus aethiopicus*** (2.7-1.9 million years ago) and ***Australopithecus garhi*** (2.6-2.5 million years ago).

Now I would like to discontinue this timeline in order to mention another information of particular importance.

Nicholas Toth and Kathy Schick published on May 9, 2018, in Azania: Archaeological Research in Africa and online study *An overview of the cognitive implications of the Oldowan Industrial Complex*.

The authors explain: "*The Oldowan Industrial Complex comprises the earliest major group of archaeological sites showing very simple stone technologies, dating back to at least 2.6-2.5 million years ago. These technologies tend to be characterized by simple core forms made of cobbles or chunks, battered precurssors, retouched flakes and unmodified stones. This industry is contemporary with the rise of the Acheulean Industrial Complex.*"

"*If the early date (3.3 million years ago) holds for Lomekwi 3, Kenia, then the hominin taxa broadly contemporaneous with this assemblage in East Africa would include:*

-Australopithecus afarensis (cranial capacity 380-430 cubic centimeters) who lived 3.9-2.9 million years ago.

-Kenyanthropus platyops (crania capacity of 450 cubic centimeters, who lived 3.5-3.2 million years ago.

-Australopithecus garhi (cranial capacity 450 cubic centimeters) and Paranthropus aethiopicus (410 cubic centimeters) who lived 2.8-2.3 million years ago."

The cranial capacity of these taxa is within the range of that seen in modern chimpanzees and bonobos.

I will not continue the author's description of hominins at various sites of this large Oldowan complex.

Philipp Gunz, Simon Neubauer, Dean Falk, Paul Tofforeau, and Zeresenay Alemseged published on April 1, 202 a study with the title Australopithecus afarensis endocasts suggest ape-like brain organization and prolonged brain growth. The study was published in Science Advances, Vol.6, issue 14, doi 10.1126/sciadv.aaz4729.

The authors explain: "*To study brain growth and organization in the hominin species Australopithecus afarensis more than 3 million years ago, we scanned eight fossil crania using conventional and synchrotron computer tomography. Contrary to previous claims, sulcal imprints reveal an ape-like brain organization and no feature derived toward humans. Similar impressions can be seen in the chimpanzee endocranium*."

At this point, I could state that **hominin's evolution, at least until 3-2.5 million years ago, did not produce brains distinct in organization from chimpanzee**. Also, their brain size was comparable.

Now, I will quote from the same article some more interesting information.

The authors further on explain: "*McGrew (1992, 2004) has documented in detail chimpanzee tool manufacture and use, as well as a range of other cultural patterns (e.g., hand-clasping). Wynn and McGrew (1989) have argued that the range of behaviors that can be inferred from Oldowan are cognitively and behaviorally comparable to those of modern apes in terms of range of tool types, their roles in subsidence behavior and patterns of manufacture. This subject was subsequently revisited (Wynn et al. 2011) with similar conclusions and citing even more types of ape tools behavior that has been documented over the intervening period. She questioned whether interpretations of Oldowan "home bases" might, in fact, have been the product of hominins with behavioral capabilities like those of chimpanzees.*"

"*It is clear that these African apes have the basic capabilities to flake stone through hard-hammer percussion and other techniques. In a comparative study of skill levels, the bonobos were given lava cobbles from 2.6 million years old conglomerates of Gona, while representatives of modern humans also flaked these cobbles (Toth et al. 2006). Although the bonobos could flake the cobbles and produce usable flakes, their cores were less heavily reduced, producing smaller flakes on average with steeper angles.* ***The Gona cores and flakes were intermediated in skill levels between those of bonobos and modern humans, but much closer to human samples***."

"*Mercader et al. (202) argued for a "Chimpanzee Stone Age."*

Hunt (2006) suggests that early stone toolmaking hominins were likely partly arboreal in their food-gathering activities and for sleeping at night and avoiding predators."

My conclusions about the documentation presented in the above paper are **that from the inception of the bipedalism (allegedly 7 million years ago) to ca. 2.5-2.0 million years ago, the African hominins were almost similar in skills, behavior, and even in cultural patterns with their contemporary chimpanzees and bonobos**. Bipedalism was combined with arboreal activities.

I will continue with the timeline of hominin's evolution.

Next species were ***Paranthropus boisei*** (2.3-1.4 mya), ***Paranthropus robustus*** (2.0-1.0 mya), ***Homo rudolfensis*** (2.3-1.9 mya) (735 cubic centimeters), ***Homo ergaster*** (2.0-1.5 mya) (850 cubic centimeters), ***Homo habilis*** (2.0-1.6 mya) (600 cubic centimeters), ***Homo georgicus*** (1.8 mya), and ***Homo erectus*** (2.0 mya-50,000 years ago) (850-1050 cubic centimeters).

Up to one million years ago there were on this list 17 hominin species (but 30+ species according to various other sources).

From two million to one million years ago the hominins brain almost doubled in size (from 400 to 735-850 cubic centimeters).

As it seems to be the case, the mentioned doubling in the brain size would be related to the evolution of the bipedalism itself that needed precise motor skills and a larger memory that would make easier tracking and predicting the location of the ground-moving animals subject to hunting. Such a memory would be very distinct from that needed by an arboreal lifestyle.

The bipedalism significantly increased the area of activity of the hominins. In the meantime, the ground lifestyle required an adaptation in sign language. Here, eventually, the vocalized-sign-language had been considerably developed.

The erect posture had a great influence in the expansion of the hominin cortex because bipedalism freed the hands, allowing them to adapt to other functions, like tool making. Simultaneously, the cortex functions must adapt to considerable leg development.

As it seems, the same erect posture influenced the shortening of the female pelvis, leading to premature births with immature brains, which contributed to increase the neoteny features in the homo genus.

All the above were positive adaptations to the new terrestrial lifestyle that truly arose between 2 to 1 million years ago, and it interpolates with doubling of the brain size compared the period 3-2 million years ago and to contemporary monkeys.

Even then, **several studies did comparative brain scanning and indicated that 2 million years ago the hominins brain organization remained like chimpanzee**.

Anthropologists define the erectus as the first “true-runner” or the first hominin with running abilities. Even though the erectus migrated out of Africa in distinct phases around 1.8-1.4 million years ago, the **running ability improvement may have occurred much later or only 1.2 million years ago**.

The average lifetime of a species is 1-10 million years. Taking the displayed duration of the species on the list, as many as 13-14 species existed in the archaeological records for a durations of 400-600,000 years that is far below the average. Hence, in the first six million years of hominin existence as bipedalists (7 to 1 million years), **there is a frequent perishing for the most identified hominin species**.

Returning to the hominins of the last one million years there are ***Homo antecessor*** (800ka-200 ka), ***Homo heidelbergensis*** (500 ka-200 ka), ***Homo neanderthalensis*** (300 ka-28 ka), and ***Homo sapiens***. Of course, there is the erectus who comes from the period of the previous million years. There is Denisovan that is not on the list.

Excepting the ***antecessor*** species, all others hominins show an even sharper reduction in the species duration (under 300-250 ka).

The **switching to bipedalism (because of a climate forcing) created the hominins, but the transition took 5-6 million years, and during this period the hominin remained not much distinct from modern monkeys (chimpanzee and bonobos) in terms of brain size, skills, tool making, and behaviors**.

As reported, the hominin bipedalism was still mixed with arboreal lifestyle, while some modern monkeys also show limited bipedalism behavior like the savanna baboons living a semi-terrestrial life.

Hominin species timeline

Bipedalism brought along a weakness that mainly manifested as a significant reduction in the overall duration of each species’ timeline.

The University of Leicester (UK) published on February 25, 2016, a study with the title *Ancient chimpanzee Adam lived over one million years ago*.

Dr. Pille Hallast from the Department of Genetics, lead author of the study, explains: "*The ancestor of Y-chromosome family tree is something called Y-chromosomal Adam. We can compare the ages of Adams between the species. For humans the age is about 200,000 years, while for gorillas it's only 100,000 years. Thanks to two chimps in our sample, Tommy and Moritz, chimpanzees have an amazing ancient Adam, who lived over 1 million years ago*."

In plain language**, the chimpanzee species has a duration of over one million years, compared to modern humans, where the genetic studies found a duration of only 200,000 years**. The chimpanzees integrate well in the average duration of a species that is 1-10 million years.

Chimpanzees are the species from which Homo sapiens originated. This study confirms that the early Homo sapiens appeared around 190 ka, but it indicates that our species presents a sharp reduction in overall duration that is 5-6 times lower than our original genetic makeup.

Such a discrepancy may be attributed to the change toward bipedalism and terrestrial lifestyle. Is this a manifestation of an inherited hominin weakness?

One should consider, as a well-adapted species, only ***Homo erectus*** (sometimes called *Homo ergaster* because of the similarities in brain size, behaviors, and geographic dispersal; but ergaster disappeared from the records 1.4-1.3 million years ago)**; erectus qualifies for the most successful hominin, but he is not the species from where Homo sapiens has arisen**.

Erectus appeared in the fossils record about 2.05 million years ago.

He was the first hominin to have relatively "human-like" proportions and was "the first runner", says paleoanthropologist Andy Herries, of La Trobe University in Melbourne, Australia. "Erectus is one of the most successful species to have ever lived" he added. Erectus made two-sided, teardrop-shaped stone hand axes, the earliest of which date back 1.7

million years. The first cranium of erectus was dated 2.04 million years ago in South Africa. Around 1.6-1.2 million years ago, erectus already was on the Jordan Valley (Palestine/Israel) in Levant. Erectus dispersed from East Africa about 2 million years ago as climate change triggered the expansion of savannas.

Over the course of about 1.75 million years, erectus expanded into Western Asia, then, into the South and Eastern Asia, crossing into what is today China and Indonesia. As the archaeological records indicate, erectus was present from northern Israel (Ubeidiya) to Caucasus (Damanisi), to China (Pekin man), and to Indonesia where Homo erectus had at 108 ka his last stand. Erectus existed for almost 2 million years. It can be noticed that H. erectus disappeared from the archaeological records probably as an outcome of the second polarity reversal of the same geomagnetic excursion that occurred around 112 years ago during the Blake event.

Similarly, later H. neanderthalensis disappeared a few years after the polarity reversal that occurred during Laschamp geomagnetic excursion.

Here, I could conclude that the species from which Homo sapiens originated were placed only in the last 500,000 years. Before that time, none of the hominin species had a significant evolution, excepting erectus who is not our ancestor.

There is a hypothesis (multiregionalism) that there was no single original home for the African hominins and no specific evolutionary path of trends.

It was found that fossils have different mosaics of trends and behaviors in different places, or even in the same archaeological place, and all dated in the same epoch.

The connections were no strait as most scientists like to describe them, because here was a great flow of genes among those numerous but small groups of foragers.

The tools were kept in a relative stability, while it showed a significant variety of tools, which were used and reused in different epochs.

I like to cite and quote Yaneer Bar-Yam with his article *Complexity Rising: From Human beings to Human Civilization, a Complexity profile*, published by EOLSS Publishers, Oxford, UK, 2002.

"*Complexity is seen as a principle that generates order. Loss of complexity shows loss of golden mean, and the vanishing of self-similarity.*"

The author explained in his paper that, within a complexity, each part (or individual) follows an essentially independent course. This independence results in an average behavior on a large scale which is simple. When the independence of component parts is reduced, the scale of behaviors is increased. **To make a large collective behavior, the parts must be correlated and not independent. When the behaviors are independent, they cancel each other in their impact on the environment. These parts (individuals) annihilate each other's actions.**

"*In order to survive, an organism's behavior must reflect in some way the nature of the environment. The organisms respond to the environment in a manner dictated by the organism's internal structure. The complexity of the response is given by the complexity of that organism's behavior.*

The probability of successful survival of an organism is a measure of the relative complexity of the environmental demands to the complexity of the organism. Organisms that are less complex than the demands of the environment have a lower probability of survival."

Here, there is the story of very many species of hominins that manifested a low probability of survival.

There is a scientifically proven relation between the common features of complex systems, like self-similarity, self-organization, power-law, and fractal dimension, where all these are intertwined together.

These thoughts on complexity explain why the **hominins before 600-500 ka were characterized by a behavior that was irrelevant to the community and its achievements**. Even then, these individuals showed complex behavior and adaptability but not significant collective behaviors. Hence**, the rise of modern behaviors is associated with the rise of collective behaviors**.

The power law (exponential growth) shows that the frequency of an occurrence of a given size is inversely proportional to the power of this size.

"***The emergency of order in complex systems is based on correlations between different levels of scales****. The organization of phenomena that belong to each level in the hierarchy rules out a preferred scale of dimension.*" (Tina Komulainen-Helsinki University of Technology).

Thus, the evolution of hominins from 6 million years ago to less than 1 million years ago, represented the struggle of those 30+ species to generate a sustainable bipedal complex system for their societies or communities, and except for Erectus, all others failed.

What failed? It was one of the common features of complex systems, self-similarity, and self-organization to power law and to fractal dimension, which all are intertwined together.

Unique brain shape in Late Hominins Clade

(the last 500,000 years)

Giorgio Manzi published (in SAGE-Hindawi Access to Research, International Journal of Evolutionary Biology, Volume 2011, Article ID582678, 11 pages) the article *Before the Emergence of Homo sapiens: Overview on the Early-to-Middle Pleistocene Fossil Record.*

The author said: "*The window between 1,000 and 500 thousand years before present appears of crucial importance, including the generation of a new and more encephalized kind of humanity, referred to by many authors as H. heidelbergensis. This species greatly diversified during the Middle Pleistocene up to the formation of new variants (i.e., incipient species) that, eventually, led to the allopatric speciation of H. neanderthalensis and H. sapiens. The special case furnished by the calvarium found near Ceprano (Italy), dated 430-385 ka, offers the perspective to investigate this matter from an original perspective.*"

H. heidelberensis existed at the crucial moment of divergence that occurred from Neanderthals to sapiens; this critical crossroad of species differentiation was dated 430-385 ka. Some researchers see it as the father of both species.

However, the article explains that the skeleton found at this site was seen as being 2/3 like H. erectus. Initially, the researchers thought that this individual was a local variation of H. erectus.

Marina Melchionna, Antonio Profico, Silvia Castiglione, Gabriele Sansalone, Carmela Serio, Mirko Di Febbraro, Lorenzo Rook, and Luca Pandolfi published in Frontiers earth Science, on July 14, 2020, the study *From Smart Apes to Human Brain Boxes. A Unique Derived Brain Shape in Late Hominins Clade.*

The authors say: "*Modern humans have larger and more globular brains when compared to other primates. Such anatomical features are further reflected in the possession of a moderately asymmetrical brain with two hemispheres apparently rotated counterclockwise and slid anteroposteriorly on one another, in what is traditionally described as Yakovlean torque. More importantly, the presence of Yakovlean torque is often advocated as the exterior manifestation of our unparalleled cognitive abilities.*"

"***Several lines of evidence indicate that not all Homo species belong to the unusual big-brained class of species, the latter being restricted to Homo heidelbergensis, Homo neanderthalensis, and Homo sapiens****. These studies point to a non-gradual process of brain increase along the hominin lineage, probably prompted by the causal association between speciation and brain size.*"

"*Our understanding of the evolution of brain shape in human lineage might be experiencing a reverse trend. Despite logical enthusiasm around early findings illustrating an exquisite human brain shape and level of brain asymmetry,* ***it has been noted that the physical brain shape in H. sapiens****, which is characterized by a strong left-occipital right-frontal asymmetry known as Yakovlean torque,* ***is present to a degree in both fossil human species and great apes****. This casts doubt on the link between brain asymmetry and properly human cognition abilities and still* *suggests that the evolution of human brain shape is best viewed as a gradual process toward exaggerated asymmetry and large size*. ***Xiang et.al. (2019) points to either a punctuational evolutionary event*** *or a shift in the rate of evolution adding variability in humans.*"

Xiang remark allows me to point more precisely to an explanation in the sense of the contribution produced by the **geophysical and geomagnetic events,** known as the geomagnetic excursions, sometimes being associated with temporal polarity reversal, and a major change in the long-term pattern of polarity reversal that occurred around 800-700 ka.

This study suggests that *"brain evolution followed a distinct path joining H. heidelberensis, H. neanderthalensis, and H. sapiens. This is entirely consistent with the archaeological records of human behavior."*

"Different studies now point in the direction that ***a clear asymmetrical brain pertains to modern species, whereas the status of austrolopiths and early homo is much less obvious****."*

"*There is a positive and significant rate shift in brain shape evolution for the clade including H. heidelbergensis, H. neanderthalensis, and H. sapiens, and a negative and significant rate shift for the Hylobatidae clade.*"

"*We demonstrate that* ***the pace of brain shape evolution shifted since the appearance of the common ancestor to H. heidelbergensis, H. neanderthalensis, and H. sapiens****. Remarkably, the rate shift coincides with archaeological evidence for the emergence of symbolic culture.*"

The shift mentioned in the above study points to a threshold in the hominin evolution placed 500,000 years ago.

An anatomical shift in the Last Clade brain shape and functions

As mentioned, many studies demonstrated that the **hominins brain shape shifted since 500,000 years ago**. This shift is positive and significant and points to a punctuational evolutionary event.

Here, I like to mention that a significant anatomical shift occurred in Homo sapiens brain starting 100,000 years ago, leading to globularization of the skull, frontoparietal development, precuneus extension, frontoparietal synthesis, and a language-ready brain. This anatomical shift allows one to consider another anatomical shift, occurring some 500,000 years ago, leading to an exaggerated asymmetry and large brain size.

As it seems, the process of encephalization comes in response to dramatic changes in the environment:

-a general cooling of the planetary climate manifested as a change in the path of glaciation (from 40 ka/ 40 ka glaciation cycles to 100 ka/ 10 ka new cycle); this switching occurred in the last one million years or less.

- a change in the polarity reversal pattern (starting 735 ka), where the reversals turn from permanent into temporary, while the geomagnetic excursions became the new geomagnetic dominance of the Earth's field.

The cooling of the climate introduced long cycles of aridity and the creation of large deserts.

The Last Clade representatives seem to adapt to such changes by a **rapid encephalization process (from around 800 ka to 500 ka), where cognition and memory became significantly affected and enlarged**. Most of this enlargement affected the cerebellum placed at the back of the skull, causing an elongated shape of the skull.

At birth, the skull of all hominins shows an elongation that continues in Neanderthals until the age of eight. In modern humans, the same elongation is present at birth but continues with a globularization process. The directionality of elongation is dictated by the needed fitness of females in the birth process. No children would have larger brains at birth because this will endanger the safety of the mother and children, too.

There is evident that an anatomically changed process has occurred 800 ka to 500 ka, and again, another anatomical change started 100 ka to 30 ka.

However, such enlargement that followed the existing directional pattern affected the Last Clade members, but also the H. erectus who developed in parallel in the same era its own encephalization.

This indicates that the <u>encephalization became a generally new manifestation of the forcing factors</u> triggered by a high C14 isotopes concentration in the terrestrial atmosphere, which stimulated ROS metabolic production and associated pulses of neurogenesis.

The neurogenesis offered the needed support in providing new waves of neurons to the enlarging and elongating brain structures. Recent studies suggest that **cerebellum may play a role in nonmotor brain functioning**. It is known that cerebellum constitutes only 10% of total brain weight, but it contains more than half of all neurons in the brain.

Nevertheless, the main role of cerebellum development was to consolidate the bipedalism and the associated functions. It strengthens the characteristics making up the terrestrial lifestyle.

Cerebellum has connections via thalamus, to many brain area involved in cognition and behavior, like the dorsolateral prefrontal cortex, the medial frontal cortex, the parietal and superior temporal areas, the anterior cingulate, and the posterior hypothalamus. It is involved in the acquisition and discrimination of sensory information.

Cerebellar size was found to be **weakly correlated with memory retention** and IQ. The weakly associations indicate that **the role of the cerebellum in cognition would be intermediated by cortical areas with which it is intimately connected**. **It becomes active in learning and word generation**.

Mounting evidence suggests that cerebellum and basal ganglia are involved in more cognitive domains like those which refer to language processing as a simple vocalization aspect.

As it appears, cerebellum determines verbal fluency expressive and receptive grammar processing, corrects expression mistakes, and may contain a host of hidden written skills. Hence, it may have a role in communication and socialization.

In the above paragraphs, I gave a description of potential roles for which the cerebellum is currently being investigated by many studies.

During the evolution stage from 600-400 ka, the development was in a relatively "primitive" phase, where the build-up of potentialities within cerebellum and associated structures may have turned fully manifested only a half a million years later during the next "grand" anatomical reorganization of the brain (100 ka-30 ka) that was called "globularization" that opposed initial elongation phase.

Recent research affirms that a particular form of Neanderthal DNA had influenced the shape of the brains (Zimmer, 2018). The research conducted by Dr. Gunz found that PHLPP1 and UBR4 genes evolved to work differently in modern humans since the version of PHLPP1 gene produces extra insulation (as myelin that helps produce a long-range communication in the cerebellum), while the UBR4 leads to a faster rate of division of the neurons in the putamen that develops a crucial motor circuitry. These genetic variations could have contributed to subtle differences in neuronal connectivity, while it may indicate how cerebellum regulates motor skills and language.

The research demonstrated that the archaic Homo sapiens (around 300-200 ka) had a brain shape and size very similar to the Neanderthals. Around 120-100 ka it started a differentiation that led to a full scale globularization in "moderns" that was accomplished only 36 ka (Zimmer, 2018).

The research found that cerebellar and parietal bulging intervened gradually, starting 120-100 ka. By comparison, the **Neanderthals** had a similar endocrinal volume, **displaying more elongated endocrania with flatter parietal and cerebellar regions**.

The research concluded that spatial variation of the whole parietal form in "modern" humans might be related to **changes in size and organization of the intraparietal sulcus** (Pereira-Pedro and Bruner, 2016), as well as in proportions of the components of the inferior parietal lobule. Also, the study found that **early (or archaic) H. sapiens had a larger cerebellum than the Neanderthals** (around 200 ka).

A recent study (13 March 2013) of Eiluned Pearce, Chris Stringer, and R.I. M. Dunbar, titled *New insights into differences in brain organization between Neanderthals and anatomically modern humans* was published by proceedings of the Royal Society B, the Oxford University and Natural History Museum, London (the UK). It discusses "*possible implications of different brain organization in terms of social cognition and consider these in the context of differing abilities to cope with fluctuating resources and cultural maintenance*."

"*We hypothesize that the similarly sized brains of Neanderthals and AMHs were organized differently for at least two reasons. First, Neanderthals had larger bodies than AMHs, and, hence, they would have required proportionally more neural matter for somatic maintenance and control. Second, Neanderthals lived at high latitudes, where they would have experienced lower light levels than tropical hominins. It requires larger eyeballs to attain the same level of visual acuity and/or sensitivity as individuals living at lower latitudes. It means that* ***if Neanderthals had larger eyes than AMHs, they would also have had larger cortices***."

"*Independent-sample t-tests found no significant difference in raw/absolute endocranial volume between AMHs and Neanderthals dated 27-75 ka. However, older AMHs dated 76-200 ka had significantly larger endocrinal volume than Neanderthals of a similar date. Neanderthals dated 27-75 ka had significantly larger endocrinal volumes than those dated 76-200 ka.*

In effect, younger Neanderthals (27-75 ka) show no increase in non-somatic/non-visual brain size compared to older Neanderthals (76-200 ka) and archaic humans. This suggests that ***later Neanderthals brains comprised a significantly larger proportion of neural tissue associated with somatic and visual function compared with the brains of contemporary AMHs***."

"***This suggests that Neanderthals and AMH brains were organized differently, and, by implication, because a greater proportion of the overall brain tissue in Neanderthals, this was invested in visual and somatic systems, and proportionally less tissue was left over for other brain areas in Neanderthals compared with AMHs***."

"*Neanderthals dated 27-75 ka were predicted to have had smaller cognitive group size than contemporary fossils AMHs. Group size is a convenient index of the cognitive ability.*"

"*Whereas AMHs appear to have concentrated neural investment in social adaptations to solve ecological problems, Neanderthals seem to have adopted an alternative strategy that involved enhanced vision coupled with retention of the physical robusticity of H. heidelberensis, but* ***not superior social cognition****. For instance, only in Neanderthals, not AMHs, does body mass and hence brain volume increased over time.*"

Another issue that was not subject to this study is "olfaction".

Several research documented that olfaction in adulthood of AMHs (dated 75-27 ka) turned diminished to only a fewer new neurons added by the neurogenesis.

This implies that AMHs (75-27 ka) brain organization had concentrated toward cognitive increases, which augmented socialization and collective behaviors with which they compensated for the evolutionarily sensorial diminishes.

Variation of C14 isotopes in the atmospheric concentration over the landmass versus the coastal areas.

Over the landmass the atmospheric concentration of C14 isotopes is mainly affected by the sink produced by vegetation that is slower in C14 absorption compared to the oceanic sink.

There is known that during a geomagnetic excursion C14 isotopes are produced in the stratosphere and driven down toward ground surface, where various sinks intervene to absorb such C14 showers.

Another case to be considered is the change in the planetary climate that intervened in the last one million years, and when the creation of desert and aride areas diminished the C14 absorption in such area.

In general, the East Africa closeness to Indian Ocean affected local variation of C14, producing a higher pulsation that occurred between production and absorption.

In the stratosphere the C14 isotopes are produced continuously by the cosmogenic radiation that penetrates the planetary weakened shield. As it is known, the geomagnetic shield weakens when the intensity of magnetic field diminishes during the excursions.

However, it intervenes a tradeoff between production of C14 and its absorption caused by the mentioned sinks, but also by the process of dissipation of C14 all over a hemisphere and between the hemispheres. This trade off makes the C14 concentration vary over several months up to two years (in on to two years the dissemination spreads in both hemispheres); thus, it produces a pulsing behavior.

Another significant fact regards the unevenness of the C14 production between the high latitudes and lower latitudes. Most geomagnetic excursions imply a flipping of the geomagnetic pole toward and up to 40 degrees parallel. In the northern hemisphere, the excursional magnetic pole travels over the Eurasia and North America landmasses.

In the Southern hemisphere the corresponding flipping takes place mostly over the oceanic areas, affecting only South and East Africa, Australia, and the southern tip of the South America.

As one can see, the ground conditions may affect to some extent the local effect of C14 isotope presence, concentration, and absorption.

Hence, the African population of East Africa had a particular C14 exposure before it migrated out of the continent. This particularity was determined by proximity of the Indian Ocean and by the aridity generated within the East African Rift.

The migrants, allegedly followed the tropical path of the Southern Coast of Asia, advancing into Southeast Asia, New Guinea, and North Australia (Sahul). Hence, this branch of migrants experienced a similar C14 pulsation like the one they had in East Africa. But at the end of the route, in New Guinea and North Australia the pulsation and the concentration of C14 significantly diminished, generating a very low radiative effect.

On the contrary, the branch of migration into Middle East and into the continental Eurasia encountered higher C14 concentrations and higher pulsations.

Even if the local concentration of C14 variation was small (probably like 10-30% difference between the mentioned zones), it may have provided a distinction in the neurogenesis production among mammals, including in the hominins.

Another aspect is migration, because it exposed the landmass path of migrants to a novel, while distinct level of exposure to the mentioned ionizing radiation (caused by the C14 isotopes).

In the meantime, the hominins who lived in Eurasia had a distinct level of exposure to ionizing radiation than those living in Africa. Such differences produced significant variations in ROS and neurogenesis levels.

There is evident that the Eurasia hominins (starting 500-400 ka) were exposed to high levels of C14 variation against the Archaic Homo sapiens, who appeared only 200 ka, and experienced a lower East African exposure until their out-of-Africa migration (70-50 ka).

It is obvious that the H. heidelberensis and H. neanderthalensis had their encephalization (as a major anatomical brain change) in Eurasia, while H. sapiens started the encephalization in Africa (200 ka) as a major anatomical brain change; the second anatomical brain change ignited in Africa (100-70 ka) but it mostly evolved out of Africa (70-27 ka).

As it seems, all three hominins were significantly affected by Blake and post-Blake geomagnetic events, which had three temporal magnetic pole reversals with a total duration of almost 4-6 ka (between 120 and 90 ka).

Such a major geomagnetic event (Blake and post-Blake) continued with Norwegian-Iceland Sea Geomagnetic excursion that manifested locally from 80 to 70 ka (as Gaotai event in NW China and Bengal event in Bangladesh) and produced an additional excursional leg 57-55 ka.

*   *   *

However, the Neanderthals are assumed to appear first in Africa (around 550-650 ka) from where they spread into higher latitude Eurasia. And probably, this is the case for H. heidelberensis, too, who lived at high latitudes of Eurasia. Even H. erectus randomly inhabited high latitudes of Eurasia.

The hominins migration out of Africa around 500 ka should be connected to climate variations in North Africa and the Arabic Peninsula, which caused a pulsation between more arid and more wet climate of this region.

The climate conditions in Eurasia were different from those in Africa and this led to certain adaptations, which also manifested in the process of encephalization, causing specific while distinct developments. This variation in brain development was related to the manifestation of various senses in distinct climates. **It primarily affected the visual and olfaction senses, but other senses, too**.

Several recent studies tried to elucidate the adaptive impact of senses on encephalization.

A brief discussion on transitional hominin species

The connection or transition between various hominin species, who lived in parallel according to the archaeological records, is not proven. There is not a clear descendance from one hominin species to another. One can say that they were similar or self-similar one to another, but here the fractal structure of the universe is the culprit but not the evolution.

However, for almost 5-6 million years the hominins were not quite distinct from contemporary monkeys, and they have the same size of brains.

As The Oldowan Industry study (that covered the period 3 to 2 million years ago) lightly concluded, "**we eventually studied the evolution of monkey or monkeys-like species" but not the true-hominin species**.

It implies that no true-hominins existed prior to 0.5 million years ago.

Changes in the glaciation pattern

Now, I would like to refer to the changes in the glaciation that occurred in the last 7 million years, and the eventual effect on hominin adaptation.

A paper of National Science Foundation with the title *Why did glacial cycles intensify a million years ago?* explains:

"*About a million years ago, something big happened to the planet. There was a major shift in the response of Earth's climate system to variations in our orbit around the Sun. That shift is called the Mid-Pleistocene Transition (MPT). Before MPT, cycles between colder glacial and warmer interglacial periods happened every 41,000 years. After glacial periods became more intense, they turned intense enough to form ice sheets in the Northern Hemisphere that last 100,000 years.*"

Overall, **the change made the glacial episodes extend from the previous 41,000 years long to 100,000 years long, while the interglacials diminished their duration from 41,000 years to ca. 10,000 years**.

It could be assumed that a significant change in the inhabitation also occurred: the hominins moved from the camping in the open-air sites toward cave dwelling.

As it appears, the change in glacial cycles left an impact on the hominin's lifestyle. Erectus was forced to migrate from central and east China toward Southeast Asia (partially in the Sahul region). In general, erectus inhabited predominantly southern Eurasia.

The other hominins show a disruption in archaeological records around one million years ago. In other words, they disappear from the records. The ***Homo antecessor*** appeared only around 800 ka, followed by new species of ***Homo heidelbergensis*** at 500 ka, and ***Homo neanderthalensis*** at 400-300 ka. These new species showed a full bipedalism adaptation, and this adaptation reflected the new survival requirements for a longer duration within the cooling episodes.

Changes in the geomagnetic pattern

Another change in the Earth's systems was the magnetic pole reversal pattern:

last reversal occurred 780 ka (called Matsuyama event). Before this last reversal, the magnetic pole changed to the opposite direction (polarity sign) every 200-300,000 years.

As it has been investigated by the geophysicists, the magnetic pole flipping, and reversals are accompanied by climate change toward cooling.

During the old pattern of reversal, it should be noted that **one hominin species disappeared from the archaeological record every second reversal**, which was the case that the magnetic pole returned (after 200-300 ka) to the polarity that existed back 400,000-600,000 years.

What could be the cause of such disappearance? No biophysical, neurological, or anthropological answer is provided, yet.

As it is documented, the geomagnetic excursions and reversals stimulate in biological systems bursts of ROS production and neurogenesis. For example, erectus brain size reflects a progressive change from 750 to 1,250 cubic centimeters. His existence overlapped in part the existence of ergaster, habilis and georgicus, but his brain size was at least 50% larger than that of the overlapping hominins. His brain was 70-80% the size of the modern humans.

Comparing erectus brain size changes at several sites in China, it could be observed an increase from 750 cubic centimeters 1.7 million years ago to 1,250 cubic centimeters 400,000 years ago.

A similar but more consistent increase in the brain size of the Neanderthals could be observed during their 400,000 years of existence.

Changes caused by the East African Rift recent evolution.

Brian Handwerk wrote the article "*To Adapt to a Changing Environment 400,000 Years Ago, Early Humans Developed New Tools and Behaviors*" published in Science on October 21, 2020.

The author explains: "***Four hundred thousand years ago, extreme environmental changes rocked the East African Rift Valley****. Fresh water periodically dried up, and vast grasslands faded away-taking with them the large grazing animals hunted by the early humans. But ecological instability didn't drive people out of the region or into extinction. Instead, it sparked them to adapt with major leaps forward in their behavior and culture. Early humans developed more sophisticated tools and weapons, expanded trade networks, and even evidenced the growth of symbolic communication.*"

"*Richard Potts, director of the Human Origins Program at the Smithsonian National Museum of Natural History, and colleagues paired a study on this matter.*"

"*By 320,000 years ago early humans had replaced first-sized stone axes with smaller, sharper, more sophisticated blades and projectile points that evidenced the Middle Stone Age technology. The same achievements begin to appear regularly in Africa during the same period. They also began to use color, chiseling red or black manganese rocks likely used to make pigments and adorn their weapons, or themselves-a practice scientists often associate with the development of symbolic thought.*"

"*The study found out that after hundreds of thousands of years of stability, dramatic shifts occurred beginning about 400,000 years ago-extreme swings occurred between wet and dry periods, lake shrunk, and new types of vegetation periodically replaced large grasslands. Geological evidence at Olorgesailie (Kenya) also show how some 400,000 years ago earthshaking tectonic activity began to reshape the region-segmenting the landscape, raising hills and cliffs, and draining huge lakes-shifts that made the area more sensitive to changes like more variable rainfall.*"

"*No fossils can be found from the key transitional period at the site because the layers that once might have held them have vanished.*"

Hence, it remains unknown what species of hominins have been forced into such dramatic but adaptive changes.

However, the blades are associated with the cutting, and cutting may lead to further assemblage of the resulting items. But here, it must be mentioned that the production of stone blades was evolving extremely slowly, taking another 240,000 years to mature in the Aurignacian era production (80-40 ka).

The previous opinions about the increased aridity in Africa showed that scientists see aridity as a background that made a forced separation between the African populations of hominins. This separation vanished only some 90,000-70,000 years ago, when a wet climate event occurred all along the East Coast of Africa (new monsoon pattern related).

Discussion

In my opinion, bipedalism required adaptation among many new neural functions, and this forced an increase in the size of the neural networks. Consequently, it occurred an increase in the brain size. Approximately one-third of the hominin genome (10,000 genes) regulate neurologic development.

As the archaeological records indicate, Homo Australopithecus brain size was 400 cubic centimeters, and gradually increased to rudolfensis to 735 and to ergaster to 850 cubic centimeters.

Erectus brain size, during its evolution, gradually increased to almost double, reaching 1,029 cubic centimeters at a site in China.

More significatively, brain size increased in Neanderthals to 1410 cubic centimeters, and in Homo sapiens it increased from 1,175 to1350 cubic centimeters.

However, the increase in the brain size did not affect the tool making, where bifaces persisted from 1.5 million years ago to 200,000 years ago.

Another stimulant was the cooking. As proved, erectus used fire and cooking 460 ka, but hearts are quite rare until 100 ka.

I may assume that the mentioned weakness referred to a negentropic imbalance, where the hominin consumption of energy was higher than the intake provided, and it affected their changes in diet and their safety against predators.

To compensate for this deficiency, hominins learned how to steal and preserve fires from natural wildfires. They used such imported fires for protection against predators, for cooking, and socialization. But this approach was characterized by randomness and intermittent.

After a long while, due to trial and error, hominins learned to produce their own fire. At this point, as some hypothesis assumed, **constant cooking generated an excess of metabolic energy that hominin biology used to develop a larger brain that fitted better the adaptation to bipedalism**. However, this possibility was infirmed by more recent studies.

Thus, brain enlargement was the most important biological adaptation for bipedality features. Equally important adaptations were those in the anatomy of the leg, hand, and digestive system.

As the archaeological research concluded, the Homo sapiens managed to engineer their own fire production, while the rest of hominins still used the fire imports. This debate is not settled, yet.

Inherently as it appears, the use of fire alone did not fully resolve an inherent weak adaptation of the bipedalists.

Adjusting behavior to changing environmental demands

Frederico Laurenco and B.J. Casey have published a study in NeurosciBiobehav Rev.2013 November, 37(900): doi:10.1016/j.neubiorev.2013.03.003.

The title of their study is *Adjusting Behavior to Changing Environmental Demands with Development*.

In their Abstract, the authors explain: "*Plasticity refers to changes in the brain that enable an organism to adapt its behaviors in the face of changing environmental demands. The evolutionary role of plasticity is to provide the cognitive flexibility to learn from experiences, to monitor the world based on learned predictions, and adjust actions when these predictions are violated. Both progressive (myelination) and regressive (synaptic pruning) brain changes support this type of adaptation. Learning capabilities and behavioral performance reflect the distinct maturational timetable of subcortical learning systems and modulatory prefrontal regions*."

"*An organism's fitness may critically depend on the ability to adjust behavior to maximize the likelihood of a successful hunt... Subcortical systems have been shown to support essential, evolutionary conserved learning processes involving reward and threat*."

"*The ability to adapt and optimize behavior varies as a function of context*."

Robert G. Bednarik of Hebei Normal University, Australia (in a paper published in January 1990 with the title *On the Cognitive Development of Hominids*) explains:

"*Human knowledge is derived from applying concept-building processes to external stimuli, sensory information, thus accumulating percepts.* ***Human knowledge tends to reinforce itself through its own products, because it is continually validated and augmented by our material and cultural achievements***." "Cultural dynamics refer to the process by which the intelligent organism alters its perceptible environment through its dialectic participation in the process shaping it."

"*The phylogenetic process of* ***cognitive evolution refers to*** *how the continual reshaping of the world view and heuristic history has prompted* ***an acceleration in the production of new conceptual and behavioral innovations, and on the role played by identifiable elements in generating cultural dynamics***."

I introduced in this section of the paper a discussion about the role of cognitive evolution of the hominins because there is undoubtedly that the gradual enlargement of their brain produced a better manipulation of cognition.

Hassa in a paper published by IPEDIA on May 14, 2020, explains: "*Cognition refers to the mental process of acquiring knowledge and understanding through thought, experience, and the senses. It involves complex mental actions such as gaining knowledge, understanding, thinking, knowing, judging, and problem solving.*" Here, problem solving is awkwardly used.

Cognition cannot directly produce a problem-solving result. An acquired knowledge could solve by itself a problem, if its complexity is at least equal with the complexity of the problem to be solved; it occurs independently of the cognitive manipulations.

Memory, which is the essential aspect of cognition, helps in storing and recovering information.

However, cognition is serially manipulated and is distinct from intelligence. In my opinion, the evolution of cognition never sufficed to generate intelligent behavior. **Intelligence represents a process of mental complexification that causes complex thoughts**. Complexification implies a combinatory process, where various free-organizing things are brought together (are combined), while the resulting complex behavior represents a novel meaning. Such a novel's meaning is intelligence.

There is one more aspect of significant importance, and it refers to **non-cognitive skills**.

David Wuepper, David Zilberman and Johannes Sauer published on Taylor Francis Online in April 20219 a study with the title *Non-cognitive skills and climate change adaptation: empirical evidence from Ghana's pineapple farmers*.

The authors explain: "*We find that those with higher non-cognitive skills are more likely to respond to the adversities of climate change by adopting climate smart technologies*."

The non-cognitive skills are often referred to as **socioemotional skills**.

They cover a wide range of abilities such as conscientiousness, perseverance, teamwork, and grit (a combination of passion and perseverance).

The researchers from the Columbia University Mailman School of Public Health, the University of Texas at Austin, and Vrije Univeriteit Amsterdam published an international study in the journal *Nature*

Genetics with the title *Noncognitive Skills, Distinct From Cognitive Abilities, Are Important to Success Across the Life* (January 2021).

One of the authors, Daniel Belsky, PhD, explains:

"*What was surprising to me about our results* ***was that these noncognitive skills contributed just as much*** *to the heritability of education attainment* ***as cognitive ability***."

"*Of the total genetic influence on educational attainment, referred as heritability,* ***cognitive abilities accounted for 43 percent and noncognitive skills accounted for 57 percent***.

Like the genetics of cognitive abilities, the genetics of noncognitive skills were related to achievements outside of schooling, including holding more prestigious jobs, earning higher incomes, and living longer. And genes associated with noncognitive skills showed relationships with these other life outcomes that were as strong or stronger than the relationships seen with cognitive genetics."

"*Noncognitive skills, defined as behaviors and abilities that are not measured by traditional IQ test but are thought to help people be more successful in schools, in their jobs, and in life generally*."

"*Our GWAS (Genome Wide Association Studies) identified signals from genes that are disproportionately active in brain tissue, similar to what has been found for the genetics of cognitive abilities*" said Margharita Malenchini from Queen Mary University of London.

"*In general, the genetics of noncognitive skills were associated with higher tolerance of risks, greater willingness to forgo immediate gratification, less health-risk behavior, and delayed fertility. The researcher also observed that noncognitive skill genetics were associated with a constellation of personality traits linked with success in relationships and at work, such as being curious and eager to learn, being more emotionally stable, and being more industrious and orderly*."

"*Our result warns us against a simplistic view of genetic variants being good or bad. The same genetic variant that predisposes someone to go further in school might also elevate their risk of developing schizophrenia or another serious mental disorder*."

Other co-author institutions to this study are King's College London, University of Zurich, University of Edinburgh, University of Amsterdam, Duke University, Stanford University, University of North Carolina at Chapel Hill. University of Michigan, and the University of Otago, New Zealand.

I cited and quoted extensively this remarkably interesting international study because it can be significantly related to hominin evolution, where the genetics had demonstrated that the **noncognitive skills are ancestral and deeply embedded in our brain for hundreds of thousands of years**.

I would say that the hominins did not need a lot of cognition capabilities because their noncognitive skills did the most of a needed job for survival.

On the other hand, **the study may suggest that noncognitive skills generated the most human behaviors long before the Industrial Revolution**. One could say that **noncognitive skills were predominant during the hominin evolution and guided their achievements**.

To conclude this discussion-section, I would say that only much later in Homo sapiens evolution **(starting 80-70,000 years ago but not in any other hominin brains) the Frontoparietal development of the brain resulted in producing a Frontoparietal Synthesis as the generator of intelligence**.

Recursiveness and recursive composition.

Recursion refers to "*a process that calls itself and provides a way to make infinite use of finite means*" (Von Humboldt, 199, Chomsky 1957). In a natural language, it refers to morphological and syntactic patterns in which **one phrase is embedded inside another of the same type**.

Here, the notion of "same type" is the fundamental feature that allows the recursion to occur.

This aspect is similar with the case of a large space (named ambient space) that has embedded a smaller space named subspace. In the mentioned case, referring to "space", the mathematicians find a **covariance** that bridges these two spaces. If the case refers to a high-dimensional space, then a so-called hyperplane can reduce the space- dimension of the ambient to the space-dimension of the subspace. The reduction becomes possible only when these two high-dimensional spaces show a **codimensionality manifested on their edges**.

Allegedly, in the brain could exist "biological hyperplanes", which progressively reduce the hierarchy of high-dimensional nonlinear signals of the unconscious into the linear signals used by conscious activity.

Hence, covariance and codimensionality are natural bridges, which allow the recursive behavior to manifest between spaces. Such features and processes are "universal," and therefore they are widespread in nature.

As it is known, recursiveness manifests in the negentropic organisms. It occurs from DNA, and RNA configurations to a complete set of behaviors. Thus, like everybody else, the **hominins inherited recursiveness**.

A recursive composition is performed stepwise (that is a processing in a series of distinct stages) compositional way. It is (approximately) like the synthesis, and in neuroscience it is called Frontoparietal Synthesis.

The employment of the fire-use in the hominin's day-to day life occurred as a simple recursive behavior.

Pyeong Whan Cho, Emily Szkudlarek and Whitney Tabor published in Frontiers in Psychology on June 8, 2016, a study with the title *Discovery of a Recursive Principle: An Artificial Grammar Investigation of Human Learning of a Counting Recursion Language.*

During their research, the authors made an experiment named *Locus Prediction* where they found that recursion acquisition occurs gradually.

"*We focus on a variant of the simplest recursive language a b, and find evidence that participants trained on two levels of structure (essentially ab and aabb) generalize to the next higher level (aaabbb) more readily that participants trained on one level of structure (ab) combined with a filter sentence; participants trained up to three levels (ab, aabb, aaabbb) generalize more readily to four levels than participants trained on two levels generalize to three. We argue that all these behaviors indicate a theory of mental representation in which recursive systems lie on a continuum of grammar systems which are organized so that grammars that produce similar behaviors are near one another, and that people learning a recursive system are navigating progressively through the space of these grammars*."

However, my interpretation of this experiment is that it was created a hierarchy of grammars, with distinct but progressive grammars. The system of grammar must be seen in a regressing order, from complex to simple ones.

Here, the simple ones are the ambient space in which are embedded more complex grammars, which are subspaces gradually decreasing in space dimensionality but increasing in complexity. Trained participants navigate between such grammars because they manifest a covariation that bridges them. Hence, the participants navigate through a hierarchy of grammars because of a covariance manifested between similar behaviors placed one near another.

However, a different recursive behavior, named **recursive composition** started to develop only 70,000 years ago. It was developed only in the Homo sapiens brain due to significant changes in his brain anatomy.

Such neuronal anatomical changes have been generated by a change in the direction of neurogenesis: it was a switch from regular neurogenesis feeding the olfaction and other senses into feeding the frontoparietal development, which switch was unique to Homo sapiens and made the precuneus to develop recursive combinations.

Geomagnetic events

Because one-third of the genome inflicts neurological system, I found that a stimulant of neurogenesis would be responsible for such a genomic evolution mostly reflected in the brain increases.

As I analyzed in some of my previous papers, the biological absorption of high concentration of atmospheric C14 isotopes generates metabolic ROS production that stimulates neurogenesis.

A natural way to generate high atmospheric concentrations of C14 isotopes is provided by geomagnetic excursions.

Here I will review the **geomagnetical events, by which in the last 400,000 years the hominins may have been affected**. In my previous articles I discussed in detail such events and their impact on hominins. Now I will present an update and a refocus of the discussion.

One of the most important developments in geomagnetism represented the scientific documentation and recognition of **polarity excursions of our planet magnetic field**.

Since bipedalism originated, there have been approximately twenty full reversals per million years, many excursions, while ca. 20-30 hominin species evolved in this time interval, but all of which are not of interest for the subject of this paper.

However, even the **geomagnetic excursions** in Lower Brunhes (since 773 ka, when the last long-term reversal occurred, to 200 ka period) have not been clearly established. But I must mention the segment they covered from 450 ka to present as potential influencers.

This subject is important because it implies the Homo Neanderthalis that appeared ca. 430 ka at the beginning of the longest interglacial on record: MIS 11 that occurred from 424 ka to 396 ka.

An Early Homo sapiens appeared ca. 160-190 ka, and its “modern” version ca. 120-110 ka. The last interglacial (MIS 6 and 5), named Eemian, is dated from 130 ka to 115-110 ka. As it seems, **the “moderns” appeared in an interglacial, like the Neanderthals**.

Most of the mentioned excursions evolved as a flip of polarity away of north-south magnetic axis, encompassing a move from North Pole to 60 to 40 degrees.

If one searches the world map, he/she will find that the mentioned flipping move occurs over the Eurasia and North America, but in the southern hemisphere, the corresponding flip goes south of Africa and Australia over the oceans, while the only southern continental mass crossed is the southern pick of South America.

There is not well established the radiative significance of the initial path of flipping before the C14 isotopes spread all over the atmosphere.

Regarding the excursion path, a notable exception is the temporal reversal of polarity, when the flipping crosses the equator, and reaches the opposite magnetic pole for a duration from a few hundred years to several thousand years. After such temporal reversal, the flipping returns to the initial polarity.

As the record shows, several disputed temporal polarity reversals occurred between 450 ka and 200 ka.

Let us see the 450,000 years of evolution of these geomagnetic events called excursions.

1.Emperor event 480-460 ka (short reversal)

2. Basura 4 excursion was 420-400 ka.

3. Levantine excursion was 370-360 ka.

4. Calabria Ridge 1 was 325-315 ka.

5. Biwa 2 was 310-292 ka.

6. Calabrian Ridge “0” was 265-255 ka.

7. Jamaica-Pringle Falls was 215-205 ka.

8. Islandic Basin was 192-189 ka.

9. Biwa 1 was 186-176 ka (recently changed to 160 ka).

10. Albuquerque 146 ka.

All these nine excursions (with two or three known reversals) influenced the hominin’s Late Clade evolution (H. heidelberensis, Neanderthals and

Denisovans) by the geomagnetic disturbance they intermittently caused for 250 ka.

Oldest Homo sapiens remains (Omo, Ethiopia) were dated 160-154 ka and more recently 195 ka (at the same site). One can speculate that **the Early Homo sapiens appeared due to the combined effect generated by the Islandic Basin and Biwa I event, which developed an intermittent disturbance of at least 25 ka**.

Of special interest is **the Blake event** that was 116-112 ka. The geophysical and geological records indicate two polarity reversals (at 116.5 ka and at 112.0 ka), while the combined reversal episodes lasted 4.5 ka. Each reversal took less than one century to be accomplished.

Also, the Fram Strait event (centered at 95 ka), renamed the **post-Blake event** (98.5 ka to 92.0 ka), indicates an abrupt reversal.

During Blake and post-Blake occurred a prominent dipole low that persisted for 30 ka.

For those who are not familiar with the Last Glacial events, I would mention that ca. 74 ka it occurred a severe cooling of the climate that was on par to LGM (Last Glacial Maximum).

Norwegian-Greenland Sea excursion was 62-60 ka. Here, some regional data from China and the Gulf of Bengal found the same event' occurrence in the interval from 80 ka to 70 ka. Also, a leg of the same excursion manifested 57 ka to 55 ka.

Laschamp event, centered 41 ka, had a controversial duration that varies from one source to another and goes from 400 years to 3.5 ka. Here, it occurred **the most studied polarity reversal** that lasted ca. 100 years. As it was surveyed, the Laschamp flipping route evolved from Asia to Western Pacific and returned over Western Africa and Western Europe.

Following Laschamp, the **Mono Lake/Auckland event** was recorded at 34 ka to 31 ka. It indicated the same level of geomagnetic disturbance as during Laschamp. **This disturbance lasted from 36 ka to 28 ka**.

This event eventually can be coupled with an unnamed event that was 28 ka to 26.5 ka, and where it was manifested a strong minimum.

Hilina Pali excursion was placed somewhere between 20 ka to 17 ka by various sources.

High atmospheric concentration of C14 isotopes and their biological effect

All these geophysical events have in common the production of nuclides, where is notable the C14-isotope for making a significant concentration in the atmosphere. As the geological records show, the C14 isotopes can have atmospheric concentrations from 10-20% to 80-100% higher than normal. The standard or normal is the level recorded before 1950.

As it is known, the C14 atmospheric concentration spreads over the entire planetary atmosphere in 1,000 to 2,000 from its inception, and before being absorbed by various environmental sinks (like oceans, vegetation).

Our civilization has experienced, during the atomic bomb atmospheric experiments (1955-1983), concentrations of C14 which were as high as double compared to the mentioned standard. And no one got any medical problems because of this occurrence. This should be noted, too.

As another example, one can cite the case of background radiation, where many people live for hundreds of generations in areas with a background-radiation that produces an ionizing effect many times higher than that allowed by the current safety regulations. Here, the exposed people do not show any adverse health effects. On the contrary, such people built up a better immunization and higher resistance against most illnesses, including cancers.

The research conclusion for the background radiation areas is that a prolonged exposure to ionizing effects stimulates higher resistance to regular illnesses.

I generated this long discussion because one can extrapolate the data to the case of **our prehistoric ancestors, who many times experienced long exposures to the effects produced by the C14 higher atmospheric concentration**.

The case of Neanderthals and Denisovans, living north of the 35 degrees north parallel, would be more distinct than that of the Homo sapiens, who before the out-of-Africa migration, lived exclusively within Africa.

The Neanderthals and Denisovans experienced at least eight geomagnetic excursions (with an undetermined number of magnetic pole reversals) before even the Homo sapiens species appeared.

Considering the out-of-Africa migration originating 70-65 ka before present, one can speculate that Homo sapiens had experienced in Africa only two reversals during Blake event and one more in the post-Blake event.

<u>The Neanderthals and Denisovans, before their extinction, experienced at least twelve polarity reversals</u>.

One can speculate that Blake and post-Blake events produced a biological shock for the African inhabitants, and such radiative effects have been much more important than those affecting the Neanderthals and Denisovans, who eventually built some resistance against the effects of such geophysical occurrences.

* * *

Beyond the influence of C14 high atmospheric concentration, there is known that some radiation, like UV, still penetrate to the ground. **Such radiation may influence the hominin entropy**.

I must explain a few concepts that characterize the new scientific thinking about entropy. A leading current concept (Ty N. E. Roach, in *Use and Abuse of Entropy in Biology*, published on 7 October 2020 in journal Entropy) explains:

"I*n many systems, **order increases as a direct result of increasing entropy**. The movement of a system toward equilibrium is determined by both the maximization of entropy and the minimization of free energy. Biological systems increase global entropy and dissipation by using free energy to create local entropy minima (i.e., building up local information and order). Entropy production is simply a byproduct due to dissipative* *processes. Biological systems minimize these quantities to perform work to ultimately surviving and reproduce."*

"***Biological systems can create and maintain localized decreases in entropy by using exogeneous sources of free energy and matter. Local decrease in entropy essentially amounts to local increase in information***.

The efficiency of living systems is a matter of engineering-informed energy flux pathways."

Brain complexity manifests in the form of an entropy enhancing process. Entropy measures the variety of possible configurations, or neural states, which can be accessed at a given moment. It describes an information processing mechanism. And it is associated with intelligence, showing that entropy relates to high information content and high intelligence.

An external energy input changes the direction of the running for the entropy flow. **Entropy flow should be the carrier of information flow** because information is a projection of entropy. Thus, such absorption increases the information flow within the cerebral and nervous systems.

* * *

There is known that **C14-isotope stimulates the ROS production that, in return, leads to neurogenesis stimulation**.

However, starting 70,000 years ago, anatomical changes in the Homo sapiens brain (like an extended surface of precuneus in the parietal region along with other changes in Frontoparietal and temporal lobes) helped combine an inherited recursiveness with the newly formed compositional approach of the neuronal network, generating the features of the **recursive composition behavior**. This change is described as the Prefrontal Synthesis (PFS) that occurs in the Lateral Prefrontal Cortex (LPFC).

After 40-50,000 years from its ignition placed at 70,000 years ago, the development of PFS in the Homo sapiens brain succeeded in generating a recursive-composition behavior, a language-ready brain, complex thoughts, and intelligent behaviors.

PFS created a distinct brain where totally new behaviors begin to manifest.

Fire use and the manmade landscape alterations

Alianda M. Cornelio, Ruben E. de Bittencourt-Navarrete, Ricardo de Bittencourt Brum, Claudio M. Queiroz, and Marcos R. Costa published in Frontiers in Neuroscience (April 2016, Volume 10, Article 167). The study title is *Human Brain Expansion during Evolution Is Independent of Fire Control and Cooking.*

The authors explain in the Abstract of their study:

"*Studies have suggested that the cooking of food could be a pre-requisite for the expansion of brain size in early hominins. However, this appealing hypothesis is only supported by a mathematical model suggesting that the increasing number of neurons in the brain would constrain body size among primates due to a limited amount of calories obtained from diets. Here,* ***we show****, by using a similar mathematical model,* ***that a tradeoff between body mass and the number of brain neurons imposed by dietary constrains during hominin evolution is unlikely****. Instead, the predictable number of neurons in the hominin brain varies much more in function of foraging efficiency than body mass. We also review archaeological data* ***to show that the expansion of the brain volume in the hominin lineage is described by a linear function independent of evidence of fire control, and therefore, thermal processing of food does not account for this phenomenon****. Finally, we report experiments in mice showing that thermal processing of meat does not increase its caloric availability in mice. Altogether,* ***our data indicate that cooking is neither sufficient nor necessary to explain hominin brain expansion****.*"

Based on the findings of the above study I will not comment on general implications of fire use. I will concentrate on the environmental implications of fire use in altering the environment that is an Anthropocene effect.

The recursive composition behavior initiated an extreme change in the Homo sapiens attitude toward the environment: the altering of small patches of landscape was intended to create more favorable conditions for survival. It represented a pre-agricultural approach.

An article published online with the title *BU Archaeologists Discover Oldest Man-Made Fire* (April 3, 2012) explains:

"*For many years it was believed that humans did not use fire until 800,000 years ago. But two Boston University archaeologists have found evidence in South Africa of man-made* ***fire dating 1.2 million years****, the earliest such discovery. The finding by Francesco Berna and Paul Goldberg substantially pushes back the date that humans laid the first kindling. Berna, the lead author, and Goldberg's research was published Monday in the online edition of Proceedings of National Academy of Sciences.*

The evidence of fire was found in Wonderwerk Cave on a high plateau in South Africa. At Wonderwerk, the location of the ash was relatively deep in the cave, more than 30 yards from the entrance, out of wildfires as well as rain and wind."

A previous study (*Fire Out Of Africa* published by ScienceDaily on October 28, 2008) refers to a research of Hebrew University of Jerusalem. The researchers discovered in Israel **a making-fire place dated 790,000 years ago**.

Another study made by University of Haifa and University of Arizona (titled *Fire at Will: The emergence of habitual fire use 350,000 years ago*, published in the Journal of Human Evolution on July 8, 2014) presents some evidence that the fire use intensified some 300-400,000 years ago. The new study found **fire use becoming habitual (regularly kindle fire) around 350,000 years ago**.

As much as we find fire use further back in time, one becomes aware that other hominins than Homo sapiens were the users. But all these fire-users hominins disappeared from the archaeological records.

Fire use was initially imported from the wildfires, and it took almost one million years for the hominins to find a way to produce their artificial type of fire. It took another 200,000 years for the fire-producing-technology to become a normal industry among most of the hominin communities.

The industry of fire use remained localized to each community or tribe and served people to cook food, to clear bones and stone, heating, and few other domestic uses, like protection against predators.

It means, the **fire-use originated from a recursive aspect or behavior,** because the hominins imported from a wildfire (consider it the ambient space) a fire for their private use (consider it the subspace).

In the end, the imported fire provided an exterior type of energy to the community. It means, the hominin species needed such exterior source of energy, while their need was exclusively domestic, helping to survive harsher environmental conditions to which they cannot adapt properly. The fire-use fulfilled an adaptive weakness. Even then, the fire-use did not prevent their extinction.

As another thing, **the fire-use did not require intelligence but only noncognitive skills**.

However, **fire-use made a fundamental difference, when it became employed to alter the landscape**, and this event took place 80-70,000 years ago near Lake Malawi in Southeast Africa. But **it remained an isolated fact until 50-30,000 years ago**.

Only during the Neolithic and Chalcolithic the fire-use become a technological source of energy. Overall, fire-use made the hominins and later humans addicted to such sources of exogeneous energy, which turned mandatory for surviving new environmental changes.

The rest of the animal kingdom survived the same environmental challenges without exogeneous help, or if they did not, they simply perished.

A later parietal development in the Homo sapiens brain has been crucial in mental modeling leading to high achievements.

Emiliano Bruner, Alexandra Battaglia-Mayer, and Roberto Caminiti published on March 24, 2022, in Brain Structure and Function a study *The parietal lobe evolution and the emergence of material culture in the human genus* (doi.org/10.1007/s00429-022-02487-w).

Authors say: "*There is consensus that the parietal lobe expanded substantially during the evolution of the genus Homo, and that this expansion is somehow associated with sophisticated capacity to use tools and manufacture the complex objects and artifacts necessary for foraging, defensive behavior, housing, and for manifold individual and collective daily activities*."

"*Object construction requires a visualspatial analysis of the available materials and a plan for assembling them through ordered movement sequences, based on the mental image of a model and/or on a physical copy-model. Such processes are absent in chimpanzees, although this species can make and use simple tools and construct simple artifacts as nets. These mental processes are only present in humans probably due to the expansion of the parietal lobe and to the emergence of hemispheric specialization across evolution*."

Here, I would like to mention that the "**object construction**" **is** what has been earlier in this text called as the **Frontoparietal Synthesis and/or recursive composition**.

"*Parietal lobe might have in object construction, object manipulation and tool use, it is worth emphasizing the core role of the posterior parietal cortex (PPC) in several visomotor functions. In fact, within the distributed parietal-frontal network, there are functional domains which are essential for accurate object manipulation, through appropriate visualmotor transformations, eye-hand coordination, hand gasping and fine control of hand force*."

"*Morphometric comparisons show that, in living humans, the precuneus, which occupies the medial wall of the superior parietal lobe (SPL), is much larger than in chimpanzee. Furthermore, in humans, the intraparietal sulcus (IPS) is particularly developed and distinct from the IPS of chimpanzees and macaques. The IPL is highly specialized in core functions belonging to human species, such as social cognition and language*."

"*A quick look at the fossil record suggests that the most noticeable change in* ***the parietal bone occurs in our own species, Homo sapiens, in which***

it is definitely much larger than in any other human taxon*. Parietal bone extension and growth are particularly* ***associated with the morphology of the underlying parietal lobe****. It is apparent that modern humans display longer and larger parietal lobe, when compared with any other extinct human species*."

"*In early humans (H. ergaster/erectus, after 2 million years ago, and H. heidelberensis, between 800-300,000 years ago) we can observe an expansion of the tempo-parietal cortex, but the dorsal parietal regions are short, flat, and depressed. In Neanderthals (say between 120,000 and 50,000 years ago) there is remarkable enlargement of the IPL, but also a slight lateral bulging of the dorso-parietal region. Early modern humans (200,000-100,000 years ago) had an overall brain morphology that was like Neanderthals, but later populations (after 100,000 years ago) displayed a pronounced bulging of the dorso-parietal region*."

"*Functional imaging (PET analysis) suggests that, beyond the parietal activation (especially the IPL), there is an important contribution of the prefrontal cortex (PFC), supposedly associated with behavioral planning and executive functions*." "*Lately, in association with the large parietal surface of modern humans, we found smaller tools, like microliths, handled with fingertips*."

"*Any given species likely employed more tool types at once, and the beginning of new technology does not match with the origin of a given species. A one-to-one association between tool typology and human species is a superficial view that should be abandoned in both scientific and dissemination contexts*."

"*Studies supporting the hypothesis that the unique ability of humans to understand causal relationships between available materials and their use as tools is dependent on the expansion and/or specialization of certain parietal areas near the IPS, while constructional abilities probably rest on the specialization of more posterior parietal and dorsolateral prefrontal (dlPFC) areas and their distributed systems*."

"*The conclusion of most studies is that conceptual processing of tools would rely on connections linking frontal and parietal areas related to action knowledge, posterior temporal, and intraparietal areas encoding abstract representations of tool knowledge*."

The extensive quotations from the above studies have the scope to demonstrate the essential role played by the development of the parietal areas of the Homo sapiens brain, starting sometimes 100,000 years ago, or according to other scientific sources around 80-70,000 years ago. This process developed simultaneously with the globularization of the skull that can be dated in parallel.

However, the parietal/precuneus/globularization combined development extended until 35,000 years ago, when a new species appeared (in my humble opinion).

Are we a new hominin species? Becoming of the Speakingman

One can say that Homo sapiens brain sharply transformed due to an anatomical while opportunistic chance that occurred just ahead of a wave of extinction where all other hominins, along with many mammals, perished 40,000 years ago.

The brains of our ancestors, the chimpanzee, have an estimated 28 billion neurons, while the gorillas have some 33.4 billion neurons.

Humans gradually evolved to current brain volume that contains around 66 billion neurons. However, the Neanderthals seem to have reached over 75 billion neurons.

Somehow accidentally, the Homo sapiens brain started around 100-70,000 years ago developing a frontoparietal section and parts of parietal lobes while diminishing the regular olfaction needs. Hence, the new neurons provided by neurogenesis were directed to the buildup of the frontoparietal and parietal structures.

Allegedly, the Homo sapiens encountered a neural architecture addition, starting 100-70,000 years ago which was largely completed 35,000 years ago.

Simon Neubauer, Jean-Jacques Hublin and Philipp Gunz in their study *The evolution of modern brain shape* , published in Science Advances on January 24, 2018, explain:

"*Brain shape, however, evolved gradually within the H. sapiens lineage, reaching present-day human variation between about 100,000 and 35,000 years ago. Present-day modern humans have globular brains and globular endocasts with steep frontal bulging parietal, and enlarged, rounded cerebellar areas. Together with small and retracted faces, this globularity characterizes the modern skull. In contrast, Neandertals and other archaic Homo individuals have anterior-posteriorly elongated endocast. Developmental globularization therefore occurs in a period of high brain growth rate and is largely driven by the brain.*"

" *Around 300,000 years ago Jebel Irhoud (Israel) fossils are not globular.*"

"*Parietal bulging is not associated with an increase of the parietal surface, but cerebellar bulging is.*"

The above statement indicates that **parietal bulging occurred by an increase in the volume of this area**.

The crania from Omo 2 (a site in Ethiopia, dated 195,000 years ago) did not have any globular shape. **The globularity started to evolve 130-100 ka, but its main development occurred from 100 ka to 35 ka. Around 35 ka, the cranial shape overlapped the present-day shape existing in humans**.

"**Parietal bulging in present day humans has been linked to large shape variation in the precuneus and is related to cognition specialization**."

"Globularization parallels behavioral modernity documented by the archaeological record, and it first manifested 50-40 ka."

Other scientists (as I described in detail in my other papers) have reduced the globularization process from 80-70,000 years ago to 30,000 years ago.

However, all previous increases in the brain volume within the hominin's clade have occurred via the archaic allometric pathway.

* * *

The energetic needs of new neural structures (frontoparietal and parietal structures) became supplemented by the achievements resulting from the landscape alteration that also paved the way to some pre-agricultural buildups (documented 50-20,000 years ago similarly in Borneo, New Guinea, and Australia) and by the manifestation of an incipient agriculture dated in the North Israel (on the shores of Sea of Galilee) around 23,000 years ago.

Here, I must repeat that frequent geomagnetic events from the epoch 120,000-4,000 years ago indirectly generated bursts of neurogenesis. Such bursts of neurogenesis participated in the buildup of new structures in the Homo sapiens brain.

Maybe the pan-last geomagnetic event of the Solovki Excursion (7,500-4,000 BCE) provided the last significant stimulation for human neurogenesis. This lengthy stimulation, which occurred during a period of 3,500 years, coincides with major human achievements, which shaped the path of the Early Antiquity era.

During this geomagnetic event, the increase of the atmospheric concentration of C14 isotopes was in the medium range but still reached some 25-35% augmented levels of C14 concentration.

Why developed the frontoparietal region in the Homo sapiens brain and not in any other hominin brains? Was it a natural accident or maybe here an anatomical or otherwise explanation?

The parietal lobe is near the upper back part of the skull and behind the frontal lobe. The parietal lobe relied heavily on many other areas of the body and of the brain to receive information. It also sends the information it processes to other parts of the brain for interpretation. In short, the **parietal lobe is a major interpreter of the sensory world around us**. It deals with touch, pain, heat, cold and tension. Thus, its development did not imply an extension of the olfaction. But it plays a role in vision and hearing, and presents the ability to judge size, shape, and distance. It interprets symbols like those originated in language.

It guides navigation and control activities.

(Anatomically, the corticostriatal connections of the parietal associations' cortices show that the rostral portion of the superior parietal lobule projects predominantly to the dorsal portion of the putamen. The caudal section of the superior parietal lobule and the cortex of the upper bank of the intraparietal sulcus have connections with the caudate nucleus and with the dorsal section of the putamen.)

The above description of the parietal functions and connections indicates the importance of this brain area when it becomes subject to highly stimulated development.

All mentioned processes that the parietal lobe hosts have been significantly or even radically enhanced and supplemented. It required a lot of additional fluxes of new neurons provided by neurogenesis.

It has been a lot of neurogenesis stimuli caused by the C14 isotopes absorption, and a lot of neurogenesis production.

Energy use connects to our future.

All mentioned changes in the Homo sapiens lifestyle gradually drove toward increases in the energetic needs, which were incrementally manifested during the 40,000 years rise of an allegedly new Homo species that we name "humans". I named this new hominin type **Homo loquens** (the **speakingman**).

Is Homo loquens a new species? However, if there is a Homo sapiens with a new brain, it would affect the body, too. Hence, having a new brain associated to a new body, what would this combination stand for?

However, the newly built neural additions fundamentally contributed to initiating a working Frontoparietal Synthesis, enhanced the recursive composition behavior, generated a coherent language and an associated intelligence. All of them started to manifest 30,000 years ago. There were too many changes which became impossible for one species, like Homo sapiens, to adjust and absorb in only 30,000 years.

This would be the threshold where the Homo sapiens' own type of complexity has collapsed, and Homo loquens began to manifest as a novel complexity arising from those collapsed parts.

Homo loquens evolution indicates a strong **addiction to energy consumption because of the novel behaviors of its developing complexity**.

Composing two or more things implies an additional energetic consumption to create a novel product. A similar additional consumption of energy is needed in the brain processing of complex thoughts or for the physical products created into the surrounding world.

As the complexity of mental composition gradually increases, the Homo loquens physical output of exponentially increasing products requires **an exogenous energy production and consumption that exponentially augments**.

<u>Is this Homo loquens also a Homo energeticus</u>? Undoubtedly, this is the case. The new type of complexity manifested in Homo loquens is much larger than that manifested in Homo sapiens. The main distinction between these two complexities is the extremely rapid increase in the number of free-organizing components, which form Homo loquens cultural complexity. All these parts develop relationships, and all relationships interact with energy consumption.

All changes and transformations I have mentioned were adaptations intended to avoid extinction. All increments in exogenous energy have the scope to avoid a catastrophic turn in entropy.

<u>We have evolved from hominins to humans by constantly increasing the species' energetic consumption that stabilizes our fundamental negentropic nature</u>. Our mental outcome is expressed by physical products, which are linearly designed to mimic a negentropic concept of stability, prediction, and control.

However, our mental processing gradually increases the exogeneous complexities we create. The exogeneous complexities' emergent behavior turns increasingly distant from any complex thought. No mental solution could fix the exogeneous immensity we have created.

We mastered from the natural fundaments a proclivity to generate our own recursion, we adapted to recursion-composition (synthesis), and we extracted the energy embedded into fire. Our intelligence originates into the emergent behavior of mental complexities created by the recursive-composition behavior and its synthesis feature.

Linearity is embedded in negentropy as an approach to life-needs that requires prediction and control.

Nevertheless, we run our linearity against the surrounding natural nonlinearity. But here, the emergent behavior of those complexities we generate, and the others created by nature, all have nonlinear solutions.

Our Artificial Intelligence ascendancy could not help transcending our built-in linearity. Our generated complexities would naturally collapse because the relationship between the self-organizing components tends to organize themselves toward more complex configurations, and in this, they require more exogenous energy. Could we provide it? And how?

Maybe quantum computing may suggest ideas about where and how to get solutions. Would we understand such nonlinear suggestions?

Part Two
The Migration

Why did prehistoric people migrate out-of-Africa?

I aim to show in this paper that Homo sapiens migration out of Africa can be correlated by some evidence suggested by neurogenesis and the Science of Complexity.

Assuming that all hominins originated in Africa, the oldest evidence of hominins in Europe dates back 700,000 to 600,000 years ago, implying that a certain migration from Africa to Europe occurred at that time.

As it seems, the Neanderthals migrated out of Africa to Europe around 350,000 years ago, becoming the dominant hominins of Europe until 40,000 years ago, when Homo sapiens appeared there, too, as a result of its out of Africa migration.

However, Jessica Tierney, an associate professor at the University of Arizona, supports the idea that "*most of our species left Africa when it was dry and not wet in northeast Africa.*"

For many years it was thought by various scholars that Homo erectus dispersed out of East Africa about 2 million years ago as climate change triggered the expansion of East Africa savanna into Southern Levant. So, in theory the hominins trailed new savannas inside and outside of Africa.

However, a new study (What Drove Homo Erectus Out of Africa, published in Science on October 19, 2021) indicates that "*Homo erectus would thrive in more than one habitat and was not limited to savannas.*"

The same study points out that "*hominins likely survived because others cared for them, assisting with the hard work of gathering, hunting, and preparing raw meat and root vegetables.*" In short "*erectus had the capacity to adjust to various environments.*"

"*Around 1.4 to 1.6 million years ago Homo erectus was occupying tropical South Asia and Indonesia.*"

Potts postulated the idea that "*in periods of rapid and sustained climatic change, only individuals with certain traits will survive, thrive, and raise children, who in turn can carry those beneficial traits, shaping human evolution. They did not know they were out of Africa. They were just going over into the next valley to see what was there.*"

Potts and Faith found that "*periods of high climate variability coincided with key milestones: the emergence of bipedalism, and the development of advanced stone tool technology, migration, and brain growth.*"

"*They were successful because they were generalists, and because they had social relationships (Belmaker).*"

Dr. Menocal remains convinced that "*the making of us was fundamentally tied to changes in the environment.*"

* * *

It is known that Homo sapiens existed in Africa in various regions, which became interconnected during wet epochs and disconnected during dry ones.

During an episode of severe cooling of the global climate, which occurred cca. seventy thousand years ago, the connections between many African regions were interrupted. On top of global climate cooling, the catastrophic eruption of Mt. Toba (Indonesia) cca. seventy-four thousand years ago generated a type of regional volcanic cooling that lasted from 1,000 to almost 5,000 years.

We should consider the Norwegian-Greenland Sea Geomagnetic Excursion effects that ignited around 78,000 years ago, extending to 70,000 years ago. Here, the specialists have identified it differently in distinct regions of the Earth. Some estimates define it as a split event that occurred 64.5 ka and again 57-54 ka. But in Hexi Corridor (China) it was defined 82.8-72.4 ka, and similarly in the Gulf of Bengal.

This excursion is thought to have triggered the global cooling that occurred around 70,000 years ago.

In the meantime, this excursion is a generator of cosmogenic radioactivity in the form of a high concentration of atmospheric C 14 isotopes. The assimilation of C 14 by feeding causes a metabolic stimulation of oxidative processes and species (ROS). This ROS, before being deactivated by antioxidative processes, generated bursts of neurogenesis in Homo sapiens.

It occurred bursts of undifferentiated neurons, followed by their gradual differentiation. Such pulses of neurogenesis seem to be at the root of the first manifestation of mental abstractions, which are associated with an incipient art developed by the Homo inhabitants isolated 70,000 years ago in a climate refuge at the peak of South Africa (Blombos cave).

The neurogenetic effect of this geomagnetic event also seems to ignite some waves of migration out of Africa. The migrants benefited from the mentioned neurogenesis bursts, generating undifferentiated stem cells, but these precursor cells' differentiation process was not fully developed. The undifferentiated bursts vastly increased the brain plasticity.

Hence, the migrants of these initial waves (70,000 years ago), and other earlier waves not discussed here (120,000-100,000 years ago), were not fully prepared to encounter new environments and various challenges. Their migration was intermittent in these early periods, while frequent local extinctions generated discontinued and interrupted settlements.

In this epoch (around 70,000 years ago), coastal Southern Asia's climate was not friendly. A dramatic change from 65,000 to 50,000 years ago occurred when the monsoon brought tropical and subtropical climate tendencies to such coastal areas.

Based on some archaeological findings, it could be determined that waves of Homo sapiens migration initially reached South India, South China, and Southeast Asia 75,000-65,000 years ago and New Guinea-Australia around 65,000-60,000 years ago. A recent archaeological finding in Laos even pushes such short-live inhabitation to 86, 0000 years ago.

Again, these migrants were not fully prepared to deal well with the areas they had to settle. I refer to two fundamental aspects: a weak cognitive development based on incomplete neural differentiation and a second weakness manifested against local pathogens and viruses.

Such a mentioned situation of weakness caused frequent extinctions among various communities or groups, drastically limiting their environmental imprint.

This case reached archaeological notoriety when one compares the age of Indonesia's archaeological sites with those of New Guinea-Australia. The Indonesian ages of sites appear much younger than the New Guinean and Australian ones. In my opinion, it implies that the initial settlers in Indonesia perished at a fast rate, and their environmental imprint became almost undetectable.

On the contrary, the settlers of Papua-New Guinea and Australia showed a more robust behavior, and their settlements succeeded the trial of time. It may be the case that these people's brains encountered a better neural differentiation that helped with a more advanced adaptation and specialization to arising local needs.

It would be very speculative to explain why the settlers in the far east of this region (New Guinea-Australia) got reasonably accommodated to their environmental constraints. At the same time, those in the west (Indonesia) did not survive local conditions, while their early chains of extinction generated a 20,000-year negative imprint in their occupation of the territory.

However, the differentiation of neurons signified an increased capability toward various adaptations. The differentiation in the neural regions in charge of cognition (behavior, memory, learning, reasoning) provides a community of neurons that assembled complexities. The emergent results of such neuronal assemblies manifested as the intelligence of those neural systems.

I would say that the **burst of neurogenesis**, when the differentiation intervened, **were the creators of human intelligence**, while such neuronal processes did not manifest in any other species.

I documented these phenomena in several of my papers recently uploaded on academia.edu (like *Universal technology of language, Language brought us the intelligence of complexification*, and *The role played by the biological absorption of C 14 isotopes in stimulating neurogenesis and biophoton production*). Hence, I would not enter these aspects in-depth here.

Assumable, the burst of neurogenesis triggered changes in the brain anatomy and neural circuitry, shown by the skull rounding process. The skull rounding incipiently manifested in the last 70,000 years ago but turned more evolved only 35,000 years ago when a language revolution occurred, showing an operational language-ready brain.

The out of Africa migrants left their continent without a language-ready brain and very primitive speaking skills. The accomplishment of the mentioned issues was brought back to Africa only later by the Asian waves of reverse migration.

The new capability of neuronal differentiation became explicitly manifested around 65,000 years ago when migration from South Africa to Eastern Africa took place. East Africa's climate turned subtropical and tropical for the first time in 100,000 years, probably extending too to the coastal areas of Southwest Asia.

* * *

The migration 65,000 years ago signifies two things. The first thing is unveiled by the Science of Complexity that shows that the African evolution of Homo sapiens acquired a complexity level of "efficiency." Such a level prevented people from adapting to constant environmental changes. This conservative behavior forced them to choose small-style habitations in stable and fertile environments with relatively stable resources.

Hence, the African inhabitants find various climates and resources refuges on Africa's territory, like fertile valleys, or fertile lands around lakes, while these people became isolated within each sanctuary.

The neurogenesis bursts stimulated higher adaptability, especially among the inhabitants of the Southern African refuge, who were challenged by a more pronounced cooling around 70,000 years ago.

East Africa's tropicalization caused by a new path in the Indian Ocean monsoon opened a window of opportunity toward better feeding and more sustainable resources.

Hence, around 65,000 years ago, the Homo of South Africa started to extend into Eastern Africa, gradually traveling 500-800 km to Kenya, Somalia, Ethiopia. The same tropicalization occurred along the Arabian Peninsula's southern coastline, making fertile coastal lands to the Indian Peninsula. Crossing Bab-El-Mandeb Strait was like crossing a large African river.

However, some scientists support the theory that during the African epochs of aridity, the populations remained isolated in more fertile refugees. When the climate improved toward wet, such isolated groups restored their mobility and restored the communication between them.

* * *

It is essential to define <u>what migration meant and how worked the mechanism that produced it</u>.

Most specialists connect the size of hominin and Homo sapiens groups to certain features of the cortex. The mental ability to maintain relationships with other individuals limits the possible group size.

Based on archaeological findings, it is admitted that African groups were in the range of a minimum of 20 and a maximum of 150 individuals.

Around 70,000 years ago, the average African size of the group was 20-30 individuals. It is known that some size variation occurred between groups, while the group size increased over time.

Migration implies a particular mental state that directly and indirectly influences the fertility rate of the community.

The most sophisticated study on Human Generation Time (HGT), dated January 2023, shows that HGT before 2,000 generations (54 ka), or before the main migration out-of-Africa, is confined to the African population and was defined at 27-28 years of age. A generation length means the average age when people have children.

This data indicates that, at that time, people inside Africa had children at the age of 27 or 28. When the migration started at 54 ka, the migrants embraced a new pattern and rapidly dropped their HGT to 20-21 years of age. This means that they made the children earlier and the number of children increased.

Another conservative approach manifested in the epoch 38-29.8 ka, but around 27 ka it diminished to 27-26 years of age, further dropping again to 21-20 years of age. There is interesting to mention that the new conservative approach from 38-30 ka corresponds with series of geomagnetic excursions in general defined as Mono Lake but thought to be in fact several distinct excursions with an eventual short-term reversal. As it seems, the excursional outcome might have produced some confusion within all populations, eventually manifesting as a sense of insecurity. Nevertheless, it may not be excluded from any other causes, like some health occurring problems.

The above-mentioned information indicates that the people in Africa before 55 ka use to have children on average at the age of 27 years old, which is seen as a conservative approach. But when they left Africa, they found the need to have children earlier, or around the age of 21-20 years old. This shows a change in mental plasticity and an increased capability to adaptations to foreign conditions encountered in the new environments they were traveling in or settling.

However, the same data might show that such adaptability did not exist before 55,000 years ago, making some early migrants (86-65 ka) perish shortly when they tried to settle in new places.

Recently, archaeologists found some incomplete evidence of hominin appearance and settlement on the eastern shores of the Americas. The dates for these findings are around 55 ka, which confirmed that this was the time when the African manifested a strong intent to travel much further than any time before, reaching beyond their original homeland.

The greening of Sahara allowed the Africans to reach again and settle on the African western coasts, like in Morocco. They were entrained in extensive fishing, probably because the local fauna was still scarce, while the sea-fish resources appeared to be very large.

However, for the Last Ice Age (LIA), it is still unknown how the path of the oceanic currents was altered, making it distinct from present. The fact that some African fishermen got driven to the eastern coasts of Americas, is indicative for such significant deviations.

Now, these Africans, 55,000 years ago, have been successful in their migration toward east, reaching New Guinea and Australia.

But, as it seems, the case of migrating west (probably as an accident) toward Americas failed to achieve the goal of an enduring settlement.

While they successfully crossed the Atlantic Ocean, when they landed the boat crews shortly perished, leaving only little evidence of their short-encounter with the Americas.

It is to mention that 50,000 years ago, the fauna of Americas was quite distinct from its African counterpart.

A few examples would be intuitive for this case. The short-faced bear weighted up to 900 kg. The American lions were the largest known cats, 25-30% larger than the African ones. Dire wolves, much bigger than the Eurasian ones, were everywhere. The jaguars were larger and more ferocious. The snakes were different, too.

Many other faunal species in the Americas were strange and extremely dangerous for the accidental African settlers.

Let us return to the African case that caused the out of the continent migration.

When Homo sapiens were in South Africa, the fertility was low. But around 65,000 years ago, that rate increased, which can be related to the mentioned bursts in neurogenesis. Better adaptability supplemented by the settlement in fertile territories increased the confidence of survival of larger communities and stimulated better reproduction rates. **The increase was not in the group size but the rate of population doubling**.

Let us see the mechanism that determines a species extension in the territory. As Professor Dunbar concluded in his extensive research on human evolution, the groups encounter a <u>doubling population split into two daughter groups</u>.

For a long time in Africa, the doubling always occurred, but it produced minimal territorial advances and tiny increases in inhabitants' number. In this case, the long length of doubling reflects a minimal rise in population. The group expansion was small, too, because, at the time, the climate refuges 70,000 years ago still had the needed resources.

For example, assumable a Paleolithic group needs a subsistence area with a radius of 125 km when the resources are on average. But let's say along a fertile valley, the site of subsistence would be reduced ten times to a radius of 12 km.

The slow Homo expansion in Africa indicates that most groups inhabited very fertile areas for very long periods. The fertility here was low because the environmental constraints did not require a backup of higher fertility.

However, 65,000 years ago, the situation dramatically changed following climate change and the bursts of neurogenesis. It changed the way this mechanism regulated the behaviors of the group.

The fertility increase made the population double in shorter intervals (around one generation). If doubling occurs every 20 years, the excess population over the stable 25 individuals must move away at 125 km from the center of the initial group's territory. Hence, in 20 years, the population increases in numbers, but also expands in the territory. In such a case, in 100 years, the people will linearly expand 625 km, in 1,000 years 6,250 km, and in 10,000 years some 62,500 km.

Considering a two-dimensional scenario: the initial settlement is A; the doubling makes people move east to B; now the doubling of A after another generation makes people move north to C; the doubling of B makes people continue to drive east to D; and so on. It shows that doubling makes people move in preferential directions, where the resources seem to fit the best. They populate the territory in a manner that is somehow uneven or discontinues.

However, the distance from Somalia to Australia, following the coastal landscape, is cca. 27,000-30,000 km and the groups from the above example would need 5,000-6,000 years to cover such a distance. In the meantime, the other daughter of the same population extends to the north, covering a similar distance. The entire globe can be populated in 12,000-15,000 years.

The pattern indicated above is a simple assumption because no information exists about Middle/Late Paleolithic fertility, mortality, and other critical data.

In the meantime, one should consider that some of the groups faced significant difficulties, which could range to extinction that produced spatial unevenness and disruption in our archaeological records.

Thus, the migration was not an actual migration process based on purpose. It occurred like a move for subsistence due to population increases.

The migrants did not know where they went or where they would go because they acted following a mechanism that implied <u>territorial expansion for subsistence</u> on whatever distance was required by the subsistence area's radius. But a radius may vary from one region to another, reflecting the availability or density of resources.

In every new areal of subsistence, <u>the group was forced to mentally adapt to the size of the problems they must solve, advancing the neuronal differentiation that increased the cognition capacity</u>. **Experiencing mental complexification and producing emergent ideas led to the first cognitive intelligence** on this planet.

The migrants also genetically adapted to novel pathogens, viruses by forming new haplogroups.

All benefits produced by the out of Africa migration were brought back to the African population by back migrations.

Magnetoreception may have contributed to ancient migration.

In a study of Joseph Kirschvink, Shin Shimojo, and Ayu Matani, published by the journal **eNeuro** on March 18, 2019, these scientists from Caltech and the neuroscientists from the Tokyo University provide the first experimental evidence about *Humans being able to detect unconsciously changes in the Earth-strength magnetic field.*

Nico and Marilyn Van Wingen, professors of Geobiology referring to the **eNeuro** study said: "*Our animal ancestry argues that geomagnetic field sensors should also be there representing not the six senses but perhaps the* 10th *of* 11th *human sense to be discovered.*"

"*During the experiment on 34 human participants "changes in their brain waves correlated with changes in the magnetic field around them. Specifically, the researchers tracked the alpha rhythm in the brain, which occurs at between 8 and 13 Hertz and is a measure of whether the brain is being engaged or is in a resting or autopilot mode. When a human brain is unengaged, the alpha power is high. When something catches its attention, consciously or unconsciously, its alpha power drops. Several other sensory stimuli like vision, hearing, and touch are known to cause abrupt drops in amplitude of alpha waves in the first few seconds after the stimulus.*"

"*The experiments showed that, in some participants, alpha power began to drop from baseline levels immediately after magnetic simulation, decreasing by as much as 60 percent over several hundred milliseconds, then recovering to baseline a few seconds after the stimulus.*"

"*The test further revealed that the brain appears to be actively processing magnetic information and rejecting signals that are not "natural".*"

"*The large individual differences we found are also intriguing with regard to human evolution and the influences of modern life.*"

"*Our results rule out electrical induction and the "quantum compass" hypotheses that have been proposed for explaining the mechanism behind magnetoreception. Instead, the* ***results indicate biological magnetite as the sensory agent for human magnetoreception***."

"*Biological magnetite has been found to exist in organisms from bacteria to humans and has been linked to the geomagnetic sense in many of them.*"

The nomadic hunter-gatherer lifestyle of our ancestors should be investigated to correlate it with an eventual magnetoreception.

Let us have a very short review on geomagnetic field variation.

First, let's see what happens with geomagnetic field during "solar-quite" (Sq) time. Such variations are of external origin caused by the currents flow in the atmosphere and ionospheric wind dynamo.

Sq current system exhibits significant solar cycle variability. Current strength can be as much as three times greater for solar maximum compared to solar minimum.

It occurs long-term seasonal and solar variability in addition to a day-to-day variability.

Field-aligned currents introduce perturbations on the order of 10-20 nanotesla to the horizontal magnetic field. Predominant annual variations are observed in the Northern Hemisphere, with the peak near the lunar solstice. Here, during solar maxima the variation becomes semi-annual. In the Southern hemisphere the variation is semi-annual.

In the Northern Hemisphere the radial focus moves polarward, generating a peak near 37.5 degrees of longitude.

The research in Sq concludes that the geomagnetic field would not induce significant longitudinal variability, and most of its variations are latitudinal and affect the nondipole field.

Analyzing the Atlantic Magnetic Anomaly, geophysicists discovered a westward drift (movement) of the geomagnetic field that seems to be constantly active in the last 1,500 years.

As it concerns the latitudinal variation, such a variation is a function of latitude: at the equator is about 10% and at 60 degrees increases to 19% or double.

As the research indicates, the rapid directional variation could be as high as 0.1 degrees of latitude per year. The distance between two degrees of latitude around the parallel of 40 degrees north is ca. 111 km.

This means the variation is ca. 11 km per year on this latitude. **In ten years of homogeneous latitudinal variation the moving distance of the nondipole grid is 111 km**.

For longitude, the distance between two meridians separated by one degree is 85 km (when computed at the crossing of the 40 degrees north parallel). For longitude, the distance diminishes toward the poles, where it is zero. Obviously, it is the largest on the equator.

The variation of a particular longitude crossing the 40 degrees north parallel is 8.5 km per year that amounts to 85 km in ten years of constant variation.

One can extrapolate the above computation to an ancient migration of hominins or humans. If the subject wants to preserve the inherited magnetic coordinates, that subject must travel in ten years 111 km on the latitudinal direction or /and 85 km on longitudinal direction.

If the subject has inherited from ancestors some well-defined magnetic coordinates, that subject had a natural tendency to search for and reach out those original coordinates.

If one generation is estimated to cover 20 years, the prehistoric subjects must travel in a lifetime at least 222 km on east-west or west-east directions, or/and 170 km on north-south or south-north directions.

In a paper titled *Why prehistoric people migrated out of Africa* I presented another mechanism that contributed to the migration (see it on my site on *academia.edu*).

In the case of coastal traveling, the grid with original coordinates may turn submerged. And then, the subjects will keep searching and moving along the coast until something stops them. Is this the way the Homo sapiens got to Sahul?

Maybe I exaggerated a bit, but such were the prehistoric cases in those zones encompassing rapid directional changes of the nondipole grid.

Examples of such rapid directional variations in the last 2000 years were found in the Northern Hemisphere for the epochs 860 AD to 960 AD, and 1450 AD to 1750 AD.

In Africa, such rapid directional variations occurred from 400 to 550 AD, 550-750 AD, and 1225 to 1550 AD.

The mentioned eras may have been periods of large human migrations.

The magnetic field contains a dipole field that is affected by the movement of the poles around the north-south axis, while significant variations occur only during the duration of the geomagnetic excursions. The dipole variations are termed "secular."

The geomagnetic field also contains a nondipole field. The research on nondipole field found that this field has continual variations, which influence the entire geomagnetic field. Most of the nondipole variations develop on duration of less than 3,000 years.

Hence, for the interest of this investigation only the non-dipole field is important. Its structure shows local poles connected by nodes. The poles move directionally subduing the nodes, and sometimes they disappear to reappear later at a different location.

As one can observe, the geomagnetic field has in the Northern Hemisphere longitudinal annual and semi-annual variations about the north-south geomagnetic axis, and latitudinal dipole variations caused by the geomagnetic pole movement that affects the inclination of the field. Such polar directional variations are significant at higher latitudes and negligible at the lower latitudes near the equator.

Because most of the world population is concentrated in the Norther Hemisphere the discussion will concentrate to the geomagnetic features of this northern zone.

As the research seems to indicate the **magnetoreception for the most animals is exclusively centered on the dipole variations**, which basically are extremely rare, excepting those variations occurring during the excursional mode of the geomagnetic field.

On the contrary, the **hominins and humans** have migrated differently, and this approach is pertinent to **a nondipole magneto sensing orientation.**

Most of the nondipole variation is latitudinal. Here should be considered also its longitudinal component and the longitudinal polar variation of the dipole field that is significantly perceived by the magnetoreception of the subjects inhabiting the areas placed at latitudes higher than 40 degrees (in the case of the Northern Hemisphere 40 degrees north). Such longitudinal variation indirectly develops, too, as a latitudinal component because the polar axis changes its latitudinal location around the magnetic pole. Hence, such variation is maximal near the magnetic pole. It gradually diminishes toward the equator, where it minimally but angularly affects the location of the magnetic equator.

Let's see how the hominins and humans have been affected by the variations of the nondipole field, which develop on time scales of less than 3,000 years.

First, I should estimate that a change in the location of nondipole local poles and associated nodes have been influenced the prehistoric movement of people. This would be so, because a nondipole type of magnetoreception would make the subjects try to reach the same nondipole magnetic characteristics they have inherited. This will cause a spatial movement or migration. If the directional variation is small their movement is small, too.

Because the "grand" migrations are of special interest, I must assume that the previously mentioned westward nondipole path of variation of the last 1,500 years varied in prehistoric times. No excursions occurred in the last 2,500 years, and one can estimate that the excursional mode has a certain influence on such variation.

One bold assumption would be that, during the Homo sapiens and hominin migrations to northeast and east of Africa (120 ka, 90-80 ka, 60-55 ka), the nondipole field encountered an eastward drift.

Obviously, geomagnetic sensing must be associated with several environmental factors, like climate changes, global ocean elevation changes, vegetal cover changes, and some other changes, which are not an immediate subject to this paper.

However, there are known several back-to-Africa migrations, which for the same reason could reflect the opposite change of the nondipole drift. Some of the inner-Africa migrations can be considered, too.

An eventual westward but Atlantic migration from Africa to Americas could also be considered because of the westward geomagnetic drift.

There is no doubt that Homo Erectus and other hominins early migration could have occurred in correlation to the same geomagnetic mechanism.

Thus, the "grand" migrations were initially an eastward and northeastwards movement. The westward change in the sense of the drift later affected the Eurasian migration, like from Asia to Europe.

The reverse changes of the geomagnetic drift potentially made the migrants pendulate latitudinally from east to west, and back to east or back to west, and longitudinally from south to north and back to south.

Time and again, this bold assumption must always be associated with many other contributing but not geomagnetic factors as previously mentioned.

Could the geological fault lines have served as a guide to ancient human migrations?

Abstract

The aim of this paper is to analyze the configuration of the faultlines and to compare such geophysical path with the geographical paths used by hominins and Homo sapiens to migrate around the world. The research found that along the faultlines develop magnetic anomalies. It would analyze the role of these anomalies in migration. The sensing of anomalies would be analyzed, too.

It is made a connection between faultlines, karstic relief, caves, seismicity, and climate.

Discussion

As it is known the basement of the terrestrial crust is crossed by multiple fault lines. How much did they affect the movement and the life of the hominins?

Let's have a concise description of such fault lines in Africa and Eurasia in the regions where there is known the hominins and Homo sapiens have traveled during their spread out-of-Africa.

As the archaeological records indicate, the East African populations have been the main contributors in the out-of-Africa migration.

The geophysical maps show a major East African Rift Valley that extends north along the Red Sea toward the Jordan Rift Valley in the Middle East. From Jordan Valley the Rift goes further north to Eastern Syria reaching Anatolia.

In Anatolia, the fault lines follow an east-west path, around central Anatolia, and connect with the fault lines crossing the Aegean Sea into Balkan Peninsula.

However, around Bab-El-Mandeb Strait, the fault lines take a diverged path going west-east along current day territories of Yemen and Oman, reaching the Ormuz Strait.

Before 10,000 years ago, the today Persian Gulf has been a valley with many rivers that was placed along the fault lines located west of Zagros Mountains of Iran. These West Zagros fault lines reach Lake Van in eastern Anatolia, where they intersect with the east-west fault lines crossing Anatolia. In the meantime, the West Zagros fault line continues up north into Caucasus region, where it intersects with an east-west fault line that ends into Crimea peninsula. This last fault line is a continuation of another fault line that crosses from the Tarim Basin (of Western China) along the Ili Valley toward Lake Balhash, and it further crosses the Kazakhstan toward South Urals, reaching Crimea.

This description of the Rift Valleys region and its diverging fault lines corresponds with the known migration path from East Africa to the Middle East to Anatolia and Balkans and to North Pontic Plaines.

However, the system of the fault lines from Hormuz Strait continues eastward to the Indian Subcontinent, where they intersect the major fault line placed south of the Himalayas that connects with the major fault lines following the Southeast Asia region into New Guinea. Undoubtedly, this system of fault lines was followed by the African migration that around 60,000 years ago has reached Sahul (the combined territories of new Guinea and Australia).

From Southeast Asia the fault lines diverge north, crossing East and Central China, Eastern Mongolia, while ending into the Baikal Rift Valley.

Moreover, from Eastern Mongolia the path of the fault lines diverges again on an east-west path that crosses North China and the Tarim Basin of Xinjiang, reaching through the Ili Valley into the Lake Balhash region of central Asia.

Here I must mention that the Eastern Mongolia/Southern Siberia combined area was the origin of many populations that migrated all along different eras of prehistory and history through the Tarim Basin rims toward Kazakhstan/Central Asia and Eastern Europe.

I must mention the fault lines crossing northern Iran that connect with Central Asia fault lines.

Also, there is important to mention an important path of fault lines existing in North Africa that connect Nile Valley to the Atlas Mountains in Morocco. This last path of fault lines was a very important artery of migration from ancient to recent times.

Why do people follow such faults lines path?

The geophysical research found that along the faults are developed positive and negative magnetic anomalies.

What are such local magnetic anomalies? The planetary magnetic field strength varies from 25,000 nT to 65,000 nT. Taken the longitudinal direction, the field strength on surface is zero near equator and maximal at the poles. From latitudinal point of view, the strength of the field at ground surface is 25,000 nT at the equator and 70,000 nT in the polar regions. In the northern hemisphere in the temperate zones, it is 50-60,000 nT.

The mentioned local magnetic anomalies are from a few nt to 100nT, as a positive or negative variation. Indeed, such a local variation is very small compared to overall planetary strength of the field, but, as the research had demonstrated, this tiny variation is well-sensed by plants and animals, greatly influencing their development and life.

For example, the plant germination and growth are amplified 4.5 times in the area with negative anomalies. This is so because the negative anomalies cause an upward movement of the water table providing the needed wetness to the plant roots.

As another fact, the faut lines are prone to higher seismicity that influences the animal senses, adapting them to protecting behaviors.

However, along a single fault line would arise a mixture of negative and positive magnetic anomalies, which show a local character.

Even when the rift valley mechanism would produce an aride episode, there will continue to develop within the negative anomaly zones a large variety of plants, which will attract a lot of associated animals (herbivores and carnivores). For example, in East Africa the rift mechanism caused an abrupt change in vegetation pattern from forests to savannas, drying out many surface water sources, but the animals have instantly adapted to such a change.

Nevertheless, there is known that a certain density of the multiple fault lines leads to desertification, like in the Tarim Basin (Takla Makan Desert), in Central Asia, Middle East, and in North Africa (Sahara Desert). But in the deserts, the oasis are those places showing a negative magnetic anomaly.

However, the deserts are subject to climate changes, encountering wet periods, when grasslands and savannahs develop, and rivers and lakes are formed.

What made hominins and humans follow such fault lines? As I mentioned, along these faults are developed negative and positive magnetic anomalies, which are sensed by plants and animals.

Did ancient humans also develop magnetic sensing? The answer (as I commented in another paper) gives a mixture of pro and contra scientific opinions.

Negative magnetic anomalies are 4-5 times more sensitive to seismic activity compared to positive anomalies. It is assumed that animals developed during a long evolution a (magnetic) sense that protects them (by seeking refuge) against some of the dangers caused by the earthquakes. Also, they communicate from one to another about the incoming seismic wave.

The same sense could have further evolved into the sensing of the underground water proximity that could help to find the hidden water springs.

But here, one could say that humans just followed the path of vegetation development as it became associated with animal habitat and behaviors.

In many fault line regions appears karstic (limestone) landscapes. It is estimated that karst covers 15% of all terrestrial surfaces. The karst landscape favors the cave development because of underground water dissolving carbonate rocks. Thus, it is estimated that no more than 5-6% of the ground surface can host karstic caves, which can be accessed from the top of the ground.

Hence, along the fault lines the migrating populations would have the chance to find a cave with a source of running water that would fit for a temporary or permanent inhabitation.

As another example, the first pre-agricultural initiatives occurred in the Rift Valleys and along the fault lines zones. Here I could mention the Lake Malawi manmade modified landscape (70-60,000 years ago), the Sea of Galilee first agricultural occurrence 23,000 years ago, and the ancient Yangtse Delta where the first rice cultivation occurred 18-16,000 years ago.

Part Three
Rising of the Speakingman

Rising of the Speakingman
(Biological contributions to the completion of the modern brain and the waves of cognition)

There is known that ionizing radiation increases the Reactive Oxygen Species (ROS). In this class of ionizing radiation enters, among other radiative sources, the gamma radiation, the gamma's outcome of nuclides in the atmosphere, like C14 isotopes but also the UV radiation.

While the UV radiation can be very harmful, damaging the DNA, the doubling of the C14 atmospheric concentration is seen as an extremely low dose of ionizing radiation with no damaging effects.

During geomagnetic excursions, the cosmogenic and galactic gamma rays penetrating the Earth's atmosphere generate nuclides, like C14 isotopes. Also, the UV radiation becomes stronger during these events. Both generate an increase in metabolic ROS that must be immediately reduced by anti-oxidative processes.

Feng He, Juan Li, Zewen Liu, Chia-Chen Chuang, Wenge Yang and Li Zuo in their study *Redox Mechanism of Reactive Oxygen Species in Exercise* (published in Frontiers in Physiology, 07 November 2016) explain:

"*Acting as signaling molecule, a physiological level of ROS is essential for normal cellular functions.. For instance, exogenous antioxidant supplements have been shown to suppress muscle contractility while the addition of H2O2 relieves such an effect, suggesting that oxidants (at low level) may be imperative in facilitating muscle contraction (Reid et al.,*

1993; Powers and Jackson, 2008). However, overproduction of ROS induced by exhaustive exercise training or other stresses, along with compromised antioxidant defenses, can lead to oxidative stress and related tissue damage (Powers et al., 2011; Zuo et al., 2012).

Interestingly, proper exercise (moderate to high intensity exercise) stimulates the adaptive responses and strengthens the endogenous antioxidant defense system to combat excessive ROS."

Zsolt Radak, Orsolya Marton, Eniko Nagy, Erika Koltai, and Sataro Goto in their study *The complex role of physical exercise and reactive oxygen species on brain* (published by Science Direct June 2013, Pages 87-93) explain:

"*The effects of exercise appear to be overly complex and could include neurogenesis via neurotrophic factors, including capillarization, decreased oxidative damage, and increased proteolytic degradation by proteasome and neprilysin. Data from our and other laboratories indicate that* ***exercise-induced modulation of ROS levels plays a role in the protein content and expression of brain-derived neurotropic factors, resulting in better function and increase neurogenesis****.*

Available data indicate that accumulation of oxidative damage impairs brain function, and exercise can attenuate the accumulation of damage causing a decline in function."

However, other studies found that **"exercise" plays a role similar to biological absorption of C14 isotopes, while both increase the neurogenesis output**.

In the transition period (100 ka to 27 ka) have been plenty of geomagnetic episodes, when the atmospheric concentration of C14 isotopes almost doubled for periods from 100 years to thousands of years. The process continued from 27 ka to 2,500 years ago with another chain of geomagnetic episodes.

Allegedly, **all such episodes of higher-than-normal levels of C14 isotopes in the atmosphere have generated bursts of neurogenesis, contributing to the shaping of the current human brain**.

Let's see how biology would explain the generation of the modern brain in humans and the novel cerebral structures.

"*Human brain development is created through continuing complex interactions of genetic and environmental influences*." (Rhoshel k. Lenroot and Jay N. Giedd, Dev Psychopatol., 2008).

Complexity of such interactions (with other genes or environmental factors) indicates how a given genetic feature is expressed.

Rhoshel K. Lenroot and Jay N. Giedd in their study *The changing impact of genes and environment on brain development during childhood and adolescence: initial findings from a neuroimaging study of pediatric twins* published by NIH Public Access in 2008 indicate:

"***Genes do not code for behavior, but for the building blocks of the cells whose interactions eventually give rise to those behaviors****; conversely,* ***the translation of environmental input into persistent behavioral changes occurs through alterations in the brain systems and even structures****. The association of a particular behavioral feature with thickness in a specific cortical area may be a clue that the mechanisms involved in the development of that cortical region are also relevant to that behavior*."

"*Results in adult populations have consistently showed* ***that variation in most brain volumes is highly driven by genetic factors****.* ***Topological features as gyrification and cortical folding have strong environmental influences****. (White, Andersen, & Nopoulos, 2002). A study of cortical gray mater density in 10 MZ and 10 DZ adult twin pairs found regional variability in heritability with the highest values in frontal and temporal areas. (Thompson, Cannon, et al., 2001).*"

"***The finding that differences in anatomy between individuals are more because of genetic influences in some brain regions and environmental influences in others indicate that the same regions that were highly heritable were also those correlated most strongly with cognitive ability****. The association between IQ and gray and white matter volumes was because of genetic rather than environmental factors.*"

"*These results demonstrate* ***that functionally relevant effects of genetic variation can be detected in brain structures, an important step toward supporting the validity of brain anatomy as an intermediate phenotype between genes and behavior****. There is a relative balance of genetic and environmental factors for specific cognitive functions. In the studies described above,* ***primary motor and sensory cortices fell within regions shaped most strongly by environmental factors****. By contrast,* ***regions of association cortex which integrate information from other brain regions and support higher cognitive activities such as language and executive function were those that more clearly showed effects of genetic variation****.*"

Noelia Urban and Francois Guillemot published in Cellular Neuroscience on 27 November 2014 the study *Neurogenesis in the embryonic and adult brain: same regulations, different roles*.

The authors say: "*The transition of proliferative and multipotent NSCs to fully differentiated neurons and glia is called neurogenesis and gliogenesis, respectively. Neurons are generated from early embryonic development until early postnatal stages, with only a few neurogenic zones remaining active in the adult (Gotz and Huttner, 2005; Ming and Song, 2011; Paridaen and Huttner, 2014). In contrast, gliogenesis starts during late embryogenesis and continues in postnatal stage, with low but widespread production of both astrocytes and oligodendrocytes also occurring throughout the adult brain (Rowitch and Kriegstein, 2010; Gallo and Deneen, 2014; Guerout et al., 2014).*"

"*The addition of new neurons to the complex circuitry of the adult brain is uncovering crucial functions for the newly generated neurons in memory and behavior (Deng et al., 2010). In particular, the integration of adult-born granule cells to the hippocampus circuitry confers an extra degree of plasticity that is crucial for the acquisition of certain types of contextual memory (Jessberger et al., 2009, Sahay et al., 2011). The hippocampal neurogenesis might be a newly evolved trait in some species, including humans, aiming to enhance adaptation to continuously changing environment (Kempermann, 2012).*"

"***The possibility to increase neurogenesis in aging mice by activating the quiescent stem cells pool is currently the focus of intense research****. In this regard, it was recently shown that* ***systemic factors from young animals can re-activate neurogenesis in aged mice*** *(Katsimpardi et al., 2014). However, disruption of quiescence signals can lead to a short-lived increase in neurogenesis, followed by a sharp decrease caused by a loss of quiescence NSCs (Ehm et al., 2010; Mira et al., 2010; Song et al., 2012).*"

This review study shows very clearly, based on multiple research, the role of neurogenesis in shaping the brain from embryonic stages into adult stages. It also values the possibility to re-activate and/or supplement neurogenesis as an anti-aging approach.

Thus, it makes evident the fact that **any stimuli that had produced an additional apport of neurogenesis had been fundamentally significant for the evolution of the hominin and human brain, especially in generating new cerebral structures**.

Sylvie Poluch and Sharon L. Juliano in their study *Fine-Tuning of Neurogenesis is essential for the Evolutionary Expansion of the Cerebral Cortex (published in Cerebral Cortex February 2015)* indicate:

"*Our results suggest* ***that fine-tuning of neurogenesis is a key evolutionary mechanism****, and that* ***changes in the timing****, rather than the sequence, of* ***neurogenetic events are essential for overall and regional cortical surface expansion****.*"

Xiangning Jiang and Jeannette Nardell in their study *Cellular and molecular introduction to brain development* (published Neurobiol Dis. 2016 Aug; 92(PtA); 3-17) indicate:

"*Human brain development starts with neurulation from ectoderm of embryo, and it takes on average, 20 to 25 years to mature. The protracted process is presented as both physical and experience-based maturation. It involves the generation of a wide variety of specialized neural and non-neural cell types.*"

"***Neurogenesis and formation of the general architecture of brain regions are largely completed at birth****, while maturation of the two principal glial cells (astrocytes and oligoden-drocytes), synaptogenesis and synapse pruning, and myelination represent postnatal brain growth (Giedd, 1999)****. Furthermore, brain constantly changes at the level of connectivity throughout the life span with environmental influences***."

However, there is known that at birth humans have an elongated skull from the front to the back that is like the brain configuration of the archaic Homo sapiens and Neanderthals. **Only after birth, the globularization process becomes active developing the frontoparietal region that makes the brain more rounded (globular)**.

I quoted several studies on neurogenesis and brain architecture, which may indicate in sum that environmental factors, like C14 isotopes, producing a stimulant effect on neurogenesis, are responsible for certain cerebral developments (primary motor and sensory cortices), while the adjacent genetics are responsible for other types of development (mostly cognitive functions and especially in the prefrontal cortex).

Most of such studies ignore a discussion on the essential role played in human brain architecture by the development of parietal region and precuneus, which region combined with frontal area is responsible for **compositional generation that shaped the language-ready brain and recursive compositions**. The same features are known in neuroscience as **Frontoparietal Synthesis** that is the tool that built our civilization in the last 27 ka.

"***Compositional generation is the ability to systematically combine known*** *(linguistic)* ***elements in novel ways****. This ability comes quite naturally in the last 27 ka for language learners, and thus appears to be a central inductive bias of the human learning system* (Chomsky, 2009, Lake, 2014, lake et al., 2019; Matthew Setzler, 2022)".

"*Combination of grounding and composition (named gSCAN) deals with combination of known inputs-either through encoding the grid world as a network, composing specialized modules based in the input command, or decoupling the subtasks of identifying a target object from that of producing an appropriate action sequence*." (Gao et al., 2020; Kuo et al., 2020, Heinze-Deml and Bouchacourt, 2020; Matthew Setzler, 2022).

"***We recruit the external world in our thinking process, which alleviates working memory and representational demands, and allows us to incrementally reason through a problem with the help of stable (but manipulative) representations***. " (Matthew Setzler, et al., 2022).

Within the parietal development the precuneus turns into an information processing center (hub), becoming vastly responsible for the recent but enormous expansion of the human consciousness.

Jonathan Bard in his study *Generation anatomical variation through mutations in networks-implications for evolution* (published in Journal of Anatomy in 2014) explains:

"*Genetic mutation leads to anatomical variation only indirectly because many proteins involved in generating anatomical structures in embryos operate cooperatively within molecular network. This includes gene-regulatory or control networks (CNs) for timing, signaling, and patterning together with the process networks (PNs) for proliferation, apoptosis, differentiation, and morphogenesis that they control. This paper argues that anatomical variation is achieved through a two-stage process**: mutation alters the output of CNs and perhaps the proliferation network, and such changed output alters the way the PNs constructs tissue**.*

*This system-biology approach has several implications: first, because networks contain **many cooperating proteins, they amplify the effect of genetic variation** so enabling mutation to generate a wider range of phenotypes than a single changed protein acting alone could. Second,, this amplification helps explain **how novel phenotypes can be produced relatively rapidly**. Third, because even organisms with **novel anatomical phenotypes derive from variants in standard networks**, there is no genetic barrier to their producing viable offspring. Network mutations cause heritable change in animal anatomy*."

How this discussion helps understand the role of C14 isotopes in stimulating the ROS and in association causing bursts of supplemental neurogenesis?

As it seems, **the bursts of neurogenesis would influence the brain genetics, producing mutations, which mutations have the role to rapidly regulate the neurogenetic output toward generating new cerebral tissue**. This process occurred 500-300 ka, when it was created an expansion of the cerebellar region. More recently, 75-27 ka, a similar process significantly extended the frontoparietal region, leading to globularity.

In both cases, new cerebral tissue was created (in the first case developed a cerebellar expansion, in the second one frontoparietal expansion developed) and its growth and growth-timing became regulated by the newly formed genes.

In the first case, the transition toward an expanded cerebellar region took between 200 ka to 100 ka. In the recent case, the transition took some 50 ka (75 to 27 ka).

However, the archaic cerebellar expansion is partially present at birth showing an elongated skull; undoubtedly, as the studies on Neanderthals skulls indicate, the rest of the cerebellar expansion occurred in the next five years after birth. The brain of the archaic Homo sapiens was remarkably similar in shape to that of the Neanderthals with an even larger cerebellum.

The recent cerebral expansion in Homo sapiens shows a significant frontoparietal growth after birth in the first one to two years. Such postnatal growth generates current globularity.

Basically, the fundamental genetic changes were consumed until 27 ka.

Even then, some smaller genetic changes seem to continually occur, because the modern brain mainly settled at 27 ka, but it continued to grow (in size) until 10 ka.

After 10 ka, the brain took a size-diminishing course.

Here there are two distinct genetic aspects: eventually many new genes have been created until 27 ka, but their expression remained dormant or was not manifested until other triggering/activating factors intervened.

As it can be observed, during our modern evolution, gradual gene activation has occurred in a cooperative manor, or in a group of several genes, causing many so-called "revolutions in doing and cognition."

As it is widely accepted, the main migration out-of-Africa occurred around 55 ka. This means that, **during the initial migration time, the modern brain was still in the transition epoch and was not fully globular. It implies that most cerebral changes have occurred in Eurasian space, while the influential factors for cerebral changes were Eurasian in nature**.

Here, it should be noted that the geomagnetic events were stronger manifested and significantly more frequent in the northern hemisphere north of the 45 parallel.

Another significant element was the norther hemisphere glaciation that did not have a corresponding equivalence on the African continent.

The glaciation manifested in the Eurasian environment with icesheets, high altitude snow and ice cover, and an extensive exposure of the continental shelf. The completion of the globularization occurred during the Last Glacial Maximum (some 25 ka). From 18-16 ka it started the deglaciation and the sea transgression.

Probably, all mentioned above factors were genetic influencers, **producing waves of genetic-collective- manifestations, or cognitive revolutions**, and they may justify a certain cranial increase.

The Holocene (after 10 ka) may have terminated the role of the environmental influencers, but now many dormant genes had a chance to show novel expressions.

As it appears, **in the course of time, the number of cognitive revolutions gradually increased, while the distance between them gradually decreased**.

* * *

Here, there is something else that may explain the creation of complex languages 30-27 ka that was followed by the mentioned revolutions in cognition.

A study recently published in the journal *Animal behaviors* by a team of researchers from Trinity College Dublin, led by Dr. Jackson, explained a phenomenon named "*critical flicker fusion*" that is based on the maximum speed of flashing lights an animal can see before it just looks like a steady, constant light.

One of the conclusions of this study is that <u>smaller animals experienced a decreased in time speed of their eye perception or the saccadic movement, similar in many respects to human speakers and listeners that allow them to distinguish more details of reality</u>. It was found to be an adaptation for survival. In contrast, larger animals process perceptual information with significantly fewer details.

Attention should be our ability to prioritize information. Eye movements (saccades) are the expression of this prioritizing pattern by the oculomotor system. <u>Attention is a prioritization state, and saccades are the motor capable of producing this state</u>.

Let me explain this process. The oculomotor system must fixate on a visual target: when the target moves, between these moves, there is no reading of the dynamics, causing a blind spot. This move between the fixations is called saccades, and it is the generator of the blind spot. This blind spot is the source of uncertainty because we do not know what happened in this time interval or latency.

The oculomotor system provides a solution: it compensates for the uncertainty of the delay with a prediction that is a probability, and it takes the form of the correction saccade. <u>**Corrections and predictions contain additional information provided from the unconscious, but also an unseen reality**</u>.

The oculomotor system gives us the illusion of the combined process of increasing speed. The increased length of latency decompresses the details of the reality, and during this stage, it makes available for cognition more of such decompressed details, which can be stored in memory and manipulated by recursive combination process that is also termed as Frontoparietal Synthesis. In the meantime the, if the overall speed of processing the visual information increases, such overall increase also embeds all the stages with extended latency and their content of the decompress details.

At increased speed, the brain would analyze the increased number of such stages, where in each stage there is a certain amount of decompressed details. Hence, **the increased overall speed would combine all the details displayed by each stage**.

<u>During human ageing, the saccade latency increases, providing more decompressed details</u>. But in the meantime, the overall speed decreases, processing lesser stages which become contracted.

Hence, the elderly capacity to see the decompressed details in each stage increased, while the <u>multiple stages contraction reduces the generalization capacity as a direct proportion to the amount of missing stages</u>.

However, the advent of a coherent language (30,000-23,000 years ago) introduced the researcher's assumption about when the language can first influence the eye movement record.

The mainstream accepted opinion is that <u>language</u> had introduced a change in cognitive state, <u>producing an additional latency of the saccades</u>. It was experimentally demonstrated in the range of 200ms.

Other researchers consider that this latency must be reduced to 100ms because it intervenes in mediation between the saccadic planning and language interpretation. They assumed that a cancellation of the pre-programmed eye movement would occur, halving the latency process.

J. B. Smaers, R. S. Rothman, D. R. Hudson, A. M. Blanoff and K. Safi along with other 18 authors have published in Science Advances on 28 April 2021 the study "*The evolution of mammalian brain size.*"

The authors found that recent **significant changes in the Earth's climate toward cooling forced many mammalians to adopt a smaller size**

body, while in most cases the brain remained the same. A bigger brain to body-size-ration can take better care of that smaller body.

Such change does not reflect an encephalization or an adaptation to better cognition. **In these smaller animals have occurred an existential (survival) eye adaptation (saccades) that allows them to distinguish more details of the environment by slowing down their visual perception**.

I could assume that a similar path was followed in the later development of Homo sapiens but not in Neanderthals and Denisovans. It was an increase in saccade/latency process in sapiens and no changes in the others.

In Homo sapiens, the mentioned change was an adaptation for surviving in changed environments and climate due to their migration out-of-Africa. **By contrast, the Neanderthals and Denisovans did not change their habitat and did not encounter an increased need for more details of their habitual environment**.

Development of human language increased the latency, allowing the brain to increase the size of the probability. This augmented latency generated changes in the oculomotor system that forced it to process differently.

The perceived visual reality is made of islands, and the saccades separate these islands**. The latency will extend the processing 150-200ms (milliseconds). The entire process produces a decompression of the original information because the same information is now 150-200ms longer**. We have a decompression of reality compared to normal visual processing. It is very tiny and occurs in all mammals.

In the case of slower perceptual speed, the flow of time seems to speed up, increasing. In the late Paleolithic and Mesolithic, time appeared to run slower, but going toward Antiquity, the flow of time became faster. Today, we have the most rapid time flow ever experienced, while we have the slowest perceptual speed yet encountered.

Speech input modifies attention states, leading to a bias toward one object, as speech directly describes or indirectly refers to. As the number of objects (details of the reality) **increased, the attention** (as a bias) **increased proportionally** to the number of objects (detail) processed.

Around forty-three thousand years ago, the language already existed among all humans, but it was a virtual static place of the mind. It was populated by nouns, pronouns, and adjectives. Because of the static nature, no significant language-related latency was generated.

Around twenty-three thousand years ago, verbs were added. The introduction of verbs and adverbs, as virtual explicates of motion content generated by conversation, was a natural result. The verbs inserted a pattern of virtual action within the virtual space of conversation that, until then, evolved as a static space of the mind with no virtual motion in it.

It was this metamorphose of the virtual static into virtual dynamic that created the modern brain.

Because of a bigger latency, humans became more visually distracted than the monkey. Human gain in language generated a decrease in their environmental skills.

The human eye favors near field objects as a matter of ecological adaptation. The resolution on such objects increases. This environmental adaptation for better description of near-field objects became significant when spoken language developed through human groups. It favored the conversation between individuals located closer to one another.

The language revolutions (70-43 ka, when the static brain was built, and 40-23 ka, when the dynamic brain was made effective) **were responsible for combined change of the speed of human perceptual processing**, by an increase in the saccadic latency of more than eleven times (but where the modern humans lost at least one-third of their Paleolithic vision accuracy).

In the last 130,000 years, 177 species of animals larger than 10 kg became extinct.

Is the enlarging of the saccadic latency a reason for an apparently longer lifespan in humans?

The researchers have compared the visual perception of a scene with speaking, writing/reading and they found the difference to be 60,000 times. In short, **this process decreased the speed of perceptual processing in humans by 60,000 times**.

Before the language revolution (30-27 ka), humans had a smaller latency, decompressing only a few details of the environment; the overall speed was also slow. This information combined gave them a limited but generalized understanding of the context. It was a contextual understanding powerful enough to assure surviving.

As language evolved, the details of the environment increased by longer latency and augmented decompression, revealing more content. But the understanding of the context becomes less generalized.

Today, content grows exponentially, doubling every year, but we are losing significant meaning from the context.

The mentioned study may help one understand **the connection between language evolution and human cognition**. Adding more details from the environment, this was the forcing factor that contributed to significant enlargements of cognition and memory.

Here, writing and reading came naturally as part of the same process that increases the details of the environment.

Considering gradual increases in the process of observing more details, which details must be processed by cognition, one can postulate that such a process produces waves of gradual increase of the cognition, and such waves accelerate as the time passes by. One can call them cognitive revolutions (in modern times, they are called **technical revolutions**).

All human discoveries are embedded in the new details unveiled from the environment by an increased level of decompaction or decompression (caused by longer latencies).

Part Four

How did evolve the Speakingman thinking?

Information transition from nonlinear of unconscious to linear of conscious and the overall evolution of consciousness

Information transition

How many space-dimensions have the reality that surrounds us?

How many of these space-dimensions are revealed by our system of perception?

Margaret Wertheim, a writer from Los Angeles, California, wrote in Aeon Essays (10 January 2018), an article with title *How many dimensions are there and what do they do to reality*, also published as *Radical Dimensions*.

She explains: *"Over the past century, the quest to describe the geometry of space has become a major project in theoretical physics, with experts from Albert Einstein onwards attempting to explain all the fundamental forces of nature as byproducts of the shape of space itself. While on the local level we are trained to think of space as having three dimensions, general relativity paints a picture of a four-dimensional universe, and in string theory it has 10 dimensions-or 11 if you take an extended version known as M-Theory. There are variations of the theory in 26 dimensions, and recently pure mathematicians have been electrified by a version describing space of 24 dimensions. But what are these dimensions?"*

"*From the perspective of mathematics,* ***a 'dimension'*** *is nothing more than another coordinate axis (another degree of freedom), which* ***ultimately becomes a purely symbolic concept not necessarily linked at all to the material world***."

"*None of these additional dimensions relate directly to our sensory experience. They are there in mathematics. It reduces physics to pure geometry.*"

"*Cosmologist Sean Carroll at Caltech said classical space isn't a fundamental part of reality's architecture. So, he says, space emerges from more basic things at the quantum level. The universe evolves in a mathematical realm with an almost infinite number of dimensions. Yet every one of them is a separate dimension in mathematical space described by quantum equations; every one new degree of freedom that the Universe has at its disposal.*"

Signe Dean, in an article "*The Human Brain Can Create Structures in Up to 11 Dimensions*", published in Frontiers of Computational Neuroscience, June 21, 2017, refers to The Blue Brain Project, a Swiss research initiative devoted to building a supercomputer-powered reconstruction of the human brain.

"*After developing their mathematical framework* (algebraic topology) *and testing it on some virtual stimuli, the team also confirmed their results on real brain tissue in rats. The researchers could discern high-dimensional geometric structures in the brain, formed by collections of tightly connected neurons (cliques) and the empty spaces (cavities) between them.*"

"*When researchers gave their virtual brain tissue a stimulus, they saw that neurons were reacting to it in a highly organized manner. It is as if the brain reacts to a stimulus by building (and) then razing a tower of multi-dimensional blocks, starting with rods (1D), then planks (2D), then cubes (3D), and then more complex geometries with 4D, 5D, etc., said one of the team mathematicians Ran Levi from Aberdeen University in Scotland.*"

The Swiss stimulation on the virtual brain found a brain's processing of up to 11 dimensions.

In my opinion, the Swiss study may provide **a mathematical hypothesis** (again, mathematical, not a real world one) **about the number of space-dimensions a brain, assumably at unconscious level, can deal with. And this number refers to 11-space-dimensions**.

What are they? Are they the 11 curled up space-dimensions assumed by the String Theory? Are they hierarchically displayed according to an embedded space-dimensionality? Maybe some of them are just a bunch of noise that may bear no meaning. On the other hand, maybe they represent encoded information, and their decoding provides hidden meanings.

How are they decoded? Is a dimensional reduction at work? Do the brain's structures employ some mental hyperplanes (as my bold assumption has conceived)?

As some studies found out, the unconscious provides unconscious thoughts. How are they related to high-dimensions? Are they borderline expressions between high and low space-dimensions?

However, only consciousness gives them a linear (three-space-dimensions) meaning we can deal with.

To confuse a little bit more this matter I will cite another study by Bertram F. Malle, professor at Brown University, titled "*How Many Dimensions of Mind Perception Really Are There?*", published by the Proceedings of the 41st Annual Meeting of the Cognitive Science Society, Montreal, 2019.

The author says: "*Four studies provide consistent evidence for three dimensions of perceived mind: Affect (A), Moral and Mental Regulation (M), and Reality Interaction. The dimensions are not simply bundles of semantically related features but capture psychological functions of the mind-to engage with its own process, with other minds, and with the social and physical world. Under some conditions, two of the tree dimensions further divide: A divides into negative and positive (social) affect, and M divides into moral cognition and social cognition.*"

Thus, here are 3- and 5-dimensional representations of human and other minds. According to the author, these structures are emergent from the interaction of other free-organizing mental structures. However, they represent the dimensions of the conscious mind that exclusively deals with a geometrically linear three-dimensional space.

Another paper attracted my highest attention because in one of my previous papers (**Universal Technology of Language**) I assumed that a hyperplane is at work in human brain and transforms nonlinear unconscious information into linear conscious. In other words, a hyperplane transforms the invisible into visible.

The paper I refer to is "*Brain Dynamics Underlying the Nonlinear Threshold for Access to Consciousness*" written by Antoine Del Cul, Sylvain Baillet, and Stanislas Dehaene. It was published by PLOS BIOLOGY online on September 25, 2007.

(doi: 10.1371/journal.pbio.0050260).

In sum, the experiments carried out during this study made the authors to conclude:

"*Our observations are most consistent with the theory of a global workplace formed by multiple distant associative area including the prefrontal, parietal, and temporal cortices. In Figure 1, we outlined the prediction derived from this model concerning the effect of decreasing masking strength. Those main prediction were:*

Initial visual activation unaffected by masking

Progressive increase in activation depth as masking strength decreases

Nonlinear divergence at a later stage

1. *Global reverberation simultaneously engages a distributed set of prefrontal, parietal, and posterior occipitotemporal regions.*
2. *All these predictions were supported by the data. Most importantly, cortical source modeling clearly indicated a strong*

> *contribution of inferior frontal and posterior parietal cortices to the observed correlate of conscious access.*
> *It supports Lamme et al.'s proposal of « localized recurrent processing*".

I will quote several paragraphs which explain the findings of this study.

"*Source analysis suggests that the underlying cerebral mechanism was a sudden activation of a distributed bilateral frontoparietal network starting about 270 ms after stimulus onset. In the preceding period (140-270 ms), the results revealed the progressive build-up, in posterior occipitotemporal and parietal areas, of a nonlinear divergence of activation as a function of SOA, which points to a dynamic nonlinear amplification as the neural correlate of masking threshold.*"

"*The seen versus not-seen network also includes a broad activation increase along the ventral occipitotemporal pathway. The time course of activation in these regions again showed a first wave (less than 270 ms) identical for seen and not-seen trials, followed by a surge of activation solely in the seen condition. The latter effect started about 270 ms after stimulus onset and lasted until about 400 ms, thus precisely coinciding with the time window where the nonlinear SOA effect was found.*"

"*Finally, **starting around 300 ms, target-evoked activation** associated with the scalp P3 quickly **expanded to a broad fronto-parieto-temporal network**. **Crucially, its activation occurred only for SOAs associated with conscious reports**.*"

In short, this study found out that **the original nonlinear signal provided by the visual perception (the not-seen) is gradually transformed (within several well-defined stages, which occur during a time frame of 300 ms), into a seen conscious reality. Such a process takes place in the fronto-parieto-temporal regions of the brain**. Thalamus and precuneus subcortical regions are significantly involved.

It is estimated that the perception is unconscious, as is all cognition. Working memory can hold only seven bits of memory into the conscious state. Thus, 95% of our goal-directed activities are executed unconsciously. All predictions are stored in the corticothalamic region, where it starts the information filtering. As it seems, ego is bound to memory and predictions.

Basal ganglia and cerebellum seem to be the structures where the nonlinear signal is processed before this information is sent to corticothalamic structures. **Unconscious is described as a nondeclarative memory system that is repressed**.

Andrea Scalabrini, Rosy Esposito, and Clara Mucci published a study in Research in Psychotherapy on March 22, 2021, with the title *Dreaming the unrepressed unconscious and beyond: repression vs. dissociation in the oneiric functioning of severe patients*.

The authors explain:

"*The capacity to dream is here proposed as a sort of enacted manifestation of emotional memories for the development of a more cohesive and symbolic vs. fragmented, diffuse and alexithymic sense of self. A recent large-scale fMRI meta-analysis in healthy subjects suggests a multi-layered nested hierarchical model of self (Qin, Wang & Northoff, 2020).*

Lately, ***the role of dreams in unconscious learning and in memory consolidation and re-consolidation has been emphasized*** *(Stickgold, Hobson, Fosse & Fosse, 2001). Dreams might be conceived as self-states that are usually accompanied by reactivation among pre-existing memories Nalbantian, 2011), resulting in new insight experiences. Other researchers consider dreams as a manifestation that searches for solutions or the best possible adaptations for the underlying conflict. Moreover,* ***dreams seem to represent a bizarre expression of affects, emotions, images, and cognition that links the deep unconscious with consciousness, excluding any form of rationality. Such a bizarreness was recently associated with the cognitive organization during waking state of abnormal psychiatric conditions like schizophrenia*** *(Scarone et al., 2008).*"

"*According to Hobson (2009) REM sleep may constitute a preconscious state that provides a virtual reality model of the world and is connected with primary conscious that can be defined as simple awareness including mainly perceptions and emotions(Vs. secondary consciousness, more related to language, metacognition and awareness)."*

"Efrat Ginot (2015) suggests that ***dreams arise from the deep unconsciousness, finding their path into meaning and consciousness and furthermore, might be considered as enacted manifestation of unconscious emotional memories and self-system****."*

The dreaming state seems to ride the same borderline between the nonlinear of unconscious toward the linear of consciousness. Her, the preconscious description refers to a space segment that seems to be placed by the transformational mechanism immediate to a completed transformation into linear consciousness. It shows mixed outcomes.

* * *

In my previous papers, I made a hypothetical case for a hyperplane type of neuronal mechanism that reduces gradually the variety of nonlinear dimensions of the frequencies of the unconscious into the three-dimensional space of the working consciousness. The first paper I quoted above seems to support my original hypothesis because it shows that the information of the unconscious visual perception goes through several stages, which cumulate a total duration of 300 ms.

This may suggest that the processed nonlinear information has various space-dimensions and implies distinct phases of dimensional reduction. The entire process occurs gradually, during 300 ms, where nonlinear information is dimensionally reduced to the linear working space of consciousness. Conscious content evolves as a gradual process of "formation" but where the transition from unconscious to conscious states grows over as a time-consuming process.

My hypothesis implies that the neuronal system is equipped with many "filters" (hyperplanes), and each filter performs one dimensional reduction at the time. Probably, it works like a cascade of reductions, and such a process requires time, namely 300 ms. These filters must homogenize the whole variety of distinct high-dimensions to one single three-dimensional space of the working consciousness.

What a neuronal filter may do to transform nonlinear high-dimensional data into a linear space?

As the specialists suggest, that neuronal structure (thought to be in the thalamus) must transform the original input data into a higher dimensional space using a nonlinear mapping. Once the map is accomplished. Neuronal structure must find a linear separating hyperplane in the space of the map. The maximal marginal hyperplane found in the space of the map corresponds to a nonlinear separating hypersurface in the original space. Such a map is a dot product, where should act a maximal separating hyperplane.

There is a constellation of cortical regions in the rostral prefrontal cortex (PFC), dorsolateral PFC (DLPFC) and superior frontal gyrus (SFG) associated with a complex perception of frequency differences among stimuli.

The experimental data suggests a nonlinear, hierarchic model, where the activity in the rostral PFC interacts with the activity in DLFPC.

Here, **the hyperplane of neural choice must have a maximum possible margin between the hyperplane and any points of the subject data-map**. This margin divides and classifies the data.

Dimensional space reduction can occur when two spaces of different dimensionality share one or more properties. It is a common factor that ranges from codimensionality to covariation, to correlation. Covariance is maximal when the bond is linear.

Hyperplane relies on existing codimensionality.

Covariance measures how much two random variables vary together within a populace of variables.

Haiping Huan in his study *Mechanisms of dimensionality reduction and decorrelation in deep neural networks* (published in arXiv, November 27, 2018) indicates that it may exist a mean-field of input dimensionality reduction in deep neural networks that also show how covariance level (redundancy) varies along the hierarchy. He proposed that a type of hyperplane is active in the brain structures, and it contributes to transforming nonlinear unconscious information (data) into linear conscient information.

Here, the dimensionality reduction results from nested nonlinear transformations of input data. Covariance is considered a tool for determining the constitutive linear structure that results from the reduction process.

Wikipedia explains: "*the difference between a subspace S and its ambient space X is known as codimension of S concerning X. Therefore, a necessary condition for S to be a hyperplane in X is for S to have codimension one in X.*"

Hence, **a hyperplane is a subspace whose dimension is one dimension less than that of its ambient space.** The research found that the pulvinar nuclei in the thalamus function is deciding which information should be inhibited, and which should be sent to further cortical areas. The thalamic and precuneus structures collaborate in selecting information. Precuneus integrate multimodal information collected from various brain regions, thereby playing an important role in mental image processing.

In humans, expansion of the parietal surface occurs during the first year of life, and such a development is absent in chimpanzees and Neanderthals. Such an expansion is associated with the expansion of precuneus.

Precuneus plays an essential role in visuospatial integration and can be associated with recent human cognitive specialization. It is the main contributor in conscious processing and may have had a significant role in recent expansion of human consciousness. Several studies found that precuneus activity correlates with self-reflection and involves mental imagery along with episodic and autobiographical memory retrieval. The **thalamus and precuneus** are not the only filters of reality, because **Striatum** also filters from a mixed nonlinear/linear outcome the most desirable linear outcome.

Striatum receives the bulk of its incoming fibres from cerebral cortex, modifies the signal, and sends it back to generate a desired action that turns out to be the most desirable linear outcome.

I mentioned these brain structures because they are the subject of dimensional transformation of the perceived data.

In my opinion, all such filtering processes must deal with initial nonlinear data that would be transformed into convenient tree-dimensional information that is the only information that the consciousness could process.

To conclude about such a dimensional reduction of the nonlinear unconscious information I would assume that each nonlinear high-dimension has a particular ambient space where it exists.

But each ambient space (larger space) has embedded a subspace that is one dimension less than the ambient. The hyperplane selects the dimension of the subspace. Such process continues until all high-dimensions are reduced to the homogeneity of one high-dimension map. Some researchers assume that such a map is the subconscious domain. Further on, the high-dimensionality of this map is reduced to the three-dimensions of the working consciousness. As the study found, the entire reduction process takes 300 ms.

As our physics describes the world, the quantum domain is at the fundament of the world, and from this domain would arise the high and low dimensions.

Such a distribution of space-dimensions reflects the entropic principle, where the high energy must decay into lower one.

Here, there must be a natural bridge that allows the energy and its corresponding space-dimensions to decay toward lower levels.

However, such a bridge or connectome seems to work similarly to a hyperplane, and it is based on features like codimensionality and covariance. Such features allow the interdimensional transition. They are embedded into the mechanism of entropy.

If this is right, then, the brain structures must instantly transit one higher dimension into a lower one. Thus, a natural hyperplane behaves with spontaneous transitions.

As it is known, when an exogenous energy is added to a physical system, it retards the entropy decaying. The biological world is a negentropic world.

As it appears, the hierarchy of space dimensions follows this entropic principle. One can retard the run of entropy, but how could one restore the initial level of energy (entropy)?

Nature tends to temporarily reorganize into complexities. The faith of a complexity depends on the capacity of such a system to bring in the exogeneous energy necessary to feed the development of the self-organizing components and their relationships. When the system fails to feed enough exogeneous energy, the system collapses or dismantle into parts.
Here, the complexity is a natural way for matter, energy, and information to temporarily reorganize. The complex system steadily grows until the energy crisis intervenes.

However, entropy is information. A complex system has more information than a simple one.

Higher dimensions have more information than the lower ones. Information has the tendency to decay, too.

If the unconscious contains high-dimensions with high information content, then, the conscious is organized to select only the lower dimensions with lower-level information content.

The same is true about the quantum level that has the highest level of content. Between the top quantum domain and our three-dimensional conscious world, located at the bottom, nature displays a wide hierarchy of dimensions and information levels.

Eventually, this is so, because only at lower levels negentropy can be created (also because here there are minimal energy requirements to temporarily complement the habitual entropy run). Complexity appears in the neighboring place of where the entropy runs high (decays).

Another significant aspect regards the fractal organization of the universe. According to such an assumption our three-dimensional world must be self-similar to the world of high-dimensions and with the quantum domain, too. On each scale of such hierarchy the space properties are different.

If one compares the reality of the unconscious with that of the conscious, the main distinction on these distinct scales is the nonlinear vs. the linear. But the nonlinear must have a lot of distinct high-dimensions with distinct space-manifestation, where each dimension has its own space properties and its own information content.

We must assume that **the space-dimensions of the unconscious are also a collection of dimensions drastically reduced from a much larger figure because our perceptions cannot access more than a fraction of a grander reality**.

In my opinion, what we currently know about quantum mechanics is only a tiny fraction of the properties it should have. It is that fraction that transcends from the fundament into the biological world and participates in its functioning.

It is known that when the free-radicals are deactivated, the process creates biophotons, and simultaneously, a quantum property, like entanglement, is activated. Dissipation also generates entanglement.

Assumably, quantum mechanics must contain most of the information that makes the universe work.

The evolution of consciousness

Robin L. Carhart-Harris, Robert Leech, Peter J. Hellyer, Murray Shanahan, Amanda Fielding, Enzo Tagliazucchi, Dante R. Chialvo and David Nutt published I frontiers in Human Neuroscience, on 03 February 2014, (doi:10.3389/fnhum.2014.00020), a study with the title *The entropic brain: a theory of conscious states informed by neuroimaging research with psychedelic drugs*.

In the abstract of that study the authors sated:

"*Entropy is a dimensionless quantity that is used for measuring uncertainty about the state of a system, but it also implies physical quantities, where high entropy is synonymous with high disorder.*

The psychedelic state is considered an exemplary of primitive or primary state of consciousness that preceded the development of modern, adult waking consciousness. The defining feature of "primary state" is elevated entropy in certain aspects of brain function, such as the repertoire of functional connectivity motifs that form and fragment across time. Indeed, since there is a great repertoire of connectivity motifs in the psychedelic state than in normal waking consciousness, this imply that primary states may exhibit "criticality", the property of being poised at a 'critical" point in a transition zone between order and disorder where certain phenomena such as power-law scaling appear. Moreover, if primary states are critical, then this suggests that entropy is suppressed in normal waking consciousness, meaning that the brain operates just below criticality. It is argued that this ***entropy suppression furnishes normal waking consciousness with a constrained quality and associated metacognitive functions, including reality-testing and self-awareness****. It is proposed that* *entry into primary states depends on a collapse of normally highly organized activity within the default-mode network (DMN) and a decoupling between the DMN and the medial temporal lobes (which are normally significantly coupled)*."

Here I must recall the concept of space and subspace, where the large space, name ambient, has a subspace. The **ambience**, in the biological world, has high entropy and is called unconscious. The **subspace behaves with negentropy, retarding the entropy, and this produces conscious behaviors**. According to this principle, all beings must have a form of consciousness.

Now, on the margin between ambient and subspace, there exists points where occurs the transition between ambient and its subspace, making it possible for the subspace to exist and connect with the ambient. Hence, the **conscious that is a suppression of entropy** (and entropy is information) **would still have on its upper margin points with high entropy**.

As the first study mentioned above describes, the altered state of consciousness seems to be located on the mentioned margin, and therefore, it behaves with some high entropy. Here it may be placed the bizarre world of dreams.

Assumably, beyond the mentioned margin, or in the depth of the ambient, the unconscious has increased levels of high of entropy, which would be much more than bizarre and would make no sense at all for a conscious mind because no translation of meanings is possible.

The mentioned "high states" in the above study, are the states on the margin, which show high entropy and which contrast with the levels of suppressed entropy.

There is interesting to mention that the studies on "intelligence" also found that intelligence **behaves with high levels of entropy**, which means that the **"intelligent" behavior develops on around a similar margin**.

In general, because entropy is fundamentally information, high entropy means more options for a system to decide for. Of course, for a conscious mind, a huge amount of options is senseless, or it generates total confusion.

However, from the point of view of the Science of Complexity, the more options a system of particles must arrange itself, the higher is the entropy, because, again, **entropy is information**. In a circumstance as that described above (ambient space/subset space), **the <u>order of a complex system is created on the margin toward the disorder of higher entropy</u>**.

Thus, all complex systems evolve on such margins and behave with high entropy. As the system evolves further on, it may reach beyond the margin area and collapse in the world of infinite possibilities.

Hence, complexity is energy dependent, needing exogeneous energy to supplement its evolving functions. If not enough energy is provided, the complex system dissipates that represents an entropy increasing.

In biological systems, it is known that dissipation produces self-organization and entanglement. When a complex system collapses, its constituent parts self-organize themselves, and they re-self-organize into a new complex system.

Several mental diseases seem to indicate that the mentally affected individuals may experience a partial switching of their brain processing toward the same marginal domain where high entropy resides.

ScienceDaily published an article (based on Elsevier information from the Institute of Clinical Medicine, University of Oslo, Norway) on August 15, 2016, with the title *Schizophrenia emerged after humans diverged from Neanderthals*. The original title of the study is "*Did schizophrenia evolve with the development of the human brain?*"

The staff writer of ScienceDaily explains:

"*This study suggests that schizophrenia is a modern development, one that emerged after humans diverged from Neanderthals*", said John Krystal, editor of Biological Psychiatry. "*It suggests that early hominids did not have this disorder*."

"*Our findings suggest that schizophrenia vulnerability rose after the divergence of modern humans from Neanderthals*" said Ole A. Andreassen" (the lead author of the Norwegian study), "*and thus supports the hypothesis* ***that schizophrenia is a by-product of the complex evolution of the human brain***."

The authors of University of Oslo study say:

"*Schizophrenia is a mental disorder that can develop in many ways, but which is generally characterized by delusions and hallucinations. Scientists observed that illness has prospered in our species throughout our documented history. Taken to its extreme, we can say that the genes for schizophrenia must be associated with success in society.*"

"The genes that make us vulnerable to schizophrenia are intricately linked to other, more favorable functions of the brain-functions that lead to significant advantages for the species, such as language and

creativity. The schizophrenia genes are linked to cognitive functions, like the ability to think and understand."

Other researchers, not involved in this study, have hypothesized that **schizophrenia must be related more precisely to the rapid evolution of language and intelligence, which both occurred only 40,000-30,000 years ago**. The way schizophrenia manifests, from light to severe forms, may be a strong indication that the brain functions ride the borderline, mentioned above, where it is reached the higher entropy domain. Here, it is the link and the similarities of illness with high achievements in life, which are above the average person.

A study from Harvard Medical School, published on January 27, 2016, in journal Nature, had as subject the **synaptic pruning in schizophrenia patients**. The researchers discovered that the C4 gene intervenes (in 1% of worldwide documented manifestation in population, but at least 5% in not manifested cases from the total population) to eliminate some connections between neurons (synapses).

The process of connecting the neurons within a network is called "*synaptic pruning*", and it develops in humans especially during adolescence. In schizophrenic patients, the C4 gene higher than normal activity (excess pruning) results in eliminating some of the synapsis in the brain, which allows the disease to develop.

However, it is well researched that, in humans, entropy **increases with age, and this process is generated by decreasing activity in several neuron networks and in excitatory conductance in synapses**. Overall, the activity in some of the synapses decreases, similarly as in the excess pruning in schizophrenia.

In both cases (schizophrenia and aging) there is an increase in entropy.

Medical literature reports that **brain entropy (BEN) changes** appear in diseases such as **schizophrenia** (Xue et al., 2019**), autism spectrum disorder** (Easson and McIntosh, 2019) (increases), and **Alzheimer** (Xue and Guo, 2018) (gradual by catastrophic decreases).

Higher value of BEN reflects more complex and higher levels of random signals of a system (Costa et al., 2005). BEN based on fMRI shows a low-high entropy contrast between neocortex and the rest of the brain, where the neocortex produces higher mental functions.

I made this discussion on **schizophrenia** because **is indicative of human consciousness evolution**.

As it seems, our consciousness very slowly evolved since 100,000 years ago with the advent in improved tool-making, and then, a light acceleration appeared 80,000-70,000 years ago, when the first artifacts appeared in archaeological records.

A significant while rapid evolution intervened 40,000-30,000 years ago when the revolution in language manifested. The language brought along intelligence and technology, which represented an out of ordinary but sharp development, implying a borderline involvement of the brain activity, where such novel activity turned difficult to control.

Even today, the mentioned difficulty persists, because the light, almost undetected and untreated forms are affecting a good segment of our society (1% or more).

I would say that most, if not all, mental diseases represent various forms of borderline rides into the domain of high entropy. A high entropy multiplies the options, multiplicating uncertainty.

Even the common depression is a manifestation of increased uncertainty that results from high entropy. Not coincidentally, many mental diseases indicate an increased expression of intelligence that also originates in high entropy.

Talking of intelligence, I must state again that a consistent, coherent intelligence, which results from the emergent behavior produced by the complexity, is generated by a process of mental complexification that ultimately means an increased entropy.

Here, the language forced the human brain to complexify, and the process generated intelligence and technology.

It is admitted that the drugs supplement with energy the brain functions, leading to borderline rides.

In the end, I would like to mention another potential borderline feature that is the perception of the dipole and nondipole grid generated by our planet magnetic field. Such a grid, magnetic in essence, can be sensed by unconscious perception, and which is suppressed by consciousness. Eventually, such an unconscious perception is not based on a particular magnetic sensing mechanism, like it is known to exist in animals. Possibly, some animals have even developed such sensing without and beyond the magnetic sensing mechanism.

The evolution of unconscious processing from monkeys to humans

Nobel laureate Daniel Kahneman in his book Thinking Fast and Slow "*characterized automatic thought processes as fast, efficient, and typically outside the realm of conscious awareness, making them devoid of deliberation or planning.*"

"***Unconscious thoughts and feelings influence*** *not only the way we perceive ourselves and the world around us but* ***also our everyday actions****. We tend to unconsciously evaluate nearly everything we come in contact with in a crude good-or-bad manner. The unconscious automatic response even translates into our basic movements, our inclination to approach or avoid an object. Our unconscious mind may not only nudge us to choose a particular option, but it may help muster the necessary motivation to actually achieve it.* ***We must accept the reality that the unconscious asserts its presence in every moment of our lives***." (John A. Bargh, January 2014).

"*Current theories of emotion suggest that stimuli are processed via an automatically engaged neural mechanism which occurs outside conscious awareness. The conclusion can be drawn that unconscious emotion processing happens all the time and has behavioral consequences.*" (Maurits van den Noort, Peggy Bosch and Kenneth Hugdahl, with the study *Non-Linear Information Processing*, performed on skin conductance and published by the Department of Cognitive Neuroscience, Bergen, Norway, 2017).

As IT research suggests, the human unconscious mind processes environmental stimuli at 20 million bits per second, and conscious mind can process only 40 bits per second. When it is associated with a particular context, a part of such an immense flow of environmental information becomes clustered and processed unconsciously. The clustered info assembles into a complexity, where it results in an emergent behavior that we call **intuition**.

Hence, the cerebral processing of the nonlinear information gives us a resolve that interprets the emergent result as "intuition." This is the only way our brain can deal with the nonlinear reality that surrounds us.

As it seems, the "intuition" occurs when the intense thinking, associated to beta brain rhythm, diminishes its frequency power, degrading into an alpha rhythm. **Various types of intuition are also associated with different frequency bands of the alpha rhythm spectrum** (8 to 12 Hertz), and some may just cross into the upper bands of theta rhythm (3 to 8 Hertz).

The Montreal Mental Health University Institute has demonstrated (2013) that "emotional regulation phase" is associated, too, with the action of the theta rhythm, while the "emotional induction phase" is associated to alpha rhythm.

In short, the emotions are regulated by the theta rhythm. The theta rhythm also supports the interface between language and memory. Speech encoding shows a cortical coupling between theta and gamma rhythms.

As we will see later in this paper, the administration of psilocybin and LSD experimental drugs leads to a diminishing of 30% in the power of alpha rhythm. According to the same research, such diminution in the **alpha power rhythm leads to so called "primary consciousness" that is assumed to be a basic working consciousness of the Homo sapiens during the Paleolithic era before the advent of language**.

According to this research modern people experience a "**secondary consciousness**" that mostly evolved in the last 40,000-10,000 years.

I hypothesize that Homo sapiens' brain activity was associated with the alpha rhythm spectrum, while coupling with higher (like beta and gamma) rhythms, and it was somehow narrowed to the unconscious processing domain.

The Paleolithic cultures had **emotional thinkers** associated with **intuitive thinkers**. Both forms of thinking resulted from unconscious processing.

Robert Sylwester in his paper *Unconscious Emotions, Conscious Feelings*, published in The Science of Learning (November 2000/Volume 58/ Number 3/ Pages 20-24) provides a significant explanation of what the emotions are.

"*Emotions, centered principally in a small set of subcortical brain systems, is our biological thermostat, and central to cognition and educational practice.*"

"***Emotion is an innate, powerful, and principally unconscious process**. It alerts us to problems but doesn't bother us with processes that don't require conscious attention.*"

"***We don't consciously choose to be emotionally aroused**, and such arousal often interferes with what we currently are doing. In effect, our emotions tell us to stop what we are currently doing and to attend to a more important challenge. That dominance is possible because far more neural fibers project from our brain's emotional centers up into the logical/rational areas than the reverse. A sudden emotional stimulus can thus easily and immediately stop an activity. It would take the time to resolve the problem before resuming the previous activity.*"

"***Our mostly unconscious body language similarly reports our current emotional state**. It is obviously useful for social species to have a way to anticipate an imminent emotion-driven behavior and to adapt to others' body language.*"

"***Emotions respond most vigorously to high-contrast information**. Emotion will merely monitor or ignore steady states and subtle changes, staying the course rather than expending cognitive energy on what isn't problematic or fluctuating. **Emotional arousal doesn't define or solve the challenge**. <u>Although emotions don't solve our problems, they can bias the direction of response</u>.*"

"***Emotions are neither positive nor negative in themselves****. Emotions evolved to alert us to specific kind of problems, so all kind of emotions are developmentally important.*"

I quoted extensively Dr. Robert Sylwester because our Paleolithic mind was mostly dealing with emotions. Hence, **individual capacity dealing with emotions was the fundamental attribute of the Paleolithic man**. Here**, the way one was approaching the variety of emotional states had defined its emotional intelligence;** such "intelligence" identifies the individual potential.

I will quote another study (Henrik Hopp, Allison S. Troy, and Iris B. Mauss, Apr. 2011) that explains further the processing of emotion regulation.

"*Unconscious processes in* ***emotion regulation describe "goal-driven" changes to any aspect of one's emotions without making a conscious decision to do so****, without paying attention to the process of regulating one's emotions, and without engaging in deliberate control. This suggests that* ***implicit valuing of emotion regulation should unconsciously and pervasively affect individuals' engagement in emotion regulation****. The effect of unconscious emotion-regulation processes depends on its interplay with conscious emotion-regulation processes.*"

Emotional intelligence is defined as a sum of qualities and attributes of an individual, including impulse control, persistence, confidence, self-motivation, empathy, trustworthiness, adaptability, and talent. As it appears, all such **qualities are implicit, meaning that they arise from the unconscious and are automatically expressed within the behaviors of an individual**. Even then, most psychologists see that some training to improve them can be done within conscious approaches.

It is considered that a gradual process of self-domestication (as an emotional regulation process) had been fundamentally influential in shaping modern emotional intelligence.

However, a negative aspect of the unconscious processing is the formation and influence of a contextual bias.

The "*pathway to mitigate unconscious bias is through understanding and developing* **emotional intelligence** *competencies. This* **can reduce or minimize the unconscious bias**." (*The implicit biases are unconscious, and we do not know their impact on our thinking and decision-making*) (David Cory, March 24, 2018).

In the last 70,000-50,000 years, anatomical changes in the Homo sapiens brain led to a gradual alteration of the exclusively highly emotional trends, causing self-domestication, and gradually implementing an **analytical (in the West) and Holistic (in the East) thinking**. The process altered emotional behavior, indicating **a gradual expansion of the linearly driven consciousness**. The advent of **language promoted more linearity and extended multifold the secondary consciousness**.

The Paleolithic mind used **intuition** as the main tool for resolving daily problems. However, intuition, in dealing with the nonlinearity of unconscious, represents and interprets the **emergent result of a perceptually contextual complexity**.

I can speculate that **emotions may result from similar contextual complexities, where the emergent result unveils potential challenges to the regular approach in doing**. Those challenges shape emotions.

There is known that human consciousness (as secondary consciousness) evolved consistently in the last 50-10,000 years ago. However, this evolution occurred on par with significant changes in the way unconscious (particularly the unconscious or implicate memory) and conscious processing have hierarchically evolved.

I will start the discussion by referring to a study on monkeys of Wenhao Dang, Russell J. Jaffe, Xue-Lian Qi, and Christos Constantinidis (*Emergence of Nonlinear Mixed Selectivity in Prefrontal Cortex after Training*) published in JNeurosci (The Journal of Neuroscience) on 1 September 2021 (DOI:https://doi.org/10.1523/JNEUROSCI.2814-20.2021).

The authors explain: "*Neurons in the PFC are typically activated by different cognitive tasks, and by different stimuli and abstract variables within these tasks. A single neuron's selectivity for a given stimulus dimension often changes depending on its context, a phenomenon known as nonlinear mixed selectivity (NMS). It has previously been hypothesized that NMS emerges because of training to perform tasks in different contexts. We tested this hypothesis directly by examining the neuronal response of different PFC areas before and after male monkeys were trained to perform different working memory tasks involving visual stimulus locations and/or shapes. We found that training induces a modest increase in the proportion of PFC neurons with NMS exclusively for spatial working memory, but not for shape information, and mostly in mid-dorsal PFC. Nonlinear mixed selectivity also displayed little modulation across either task complexity or correct performance*."

"*Individual PFC neurons typically encode more than one variable in their activity, and the exact variables encoded are task dependent. A portion of neurons exhibit nonlinear mixed selectivity (NMS) for different variables, which means that their response to the combination of variables cannot be predicted by the linear summation of their individual response. It is therefore possible that NMS may manifest exclusively in a limited set of PFC subdivisions or, alternatively, the NMS emerges exclusively after training to perform specific types of cognitive tasks*."

"*NMS is thought to be critical in that respect, as it allows efficient representation of flexible, arbitrary combination of variables. Consistent with this data, increased dimensionality in NMS has been highlighted as a potential means in increasing efficiency of the WM (working memory) task performance, while a dimensional collapse characterizes task errors. NMS is assumed to emerge with training in complex tasks that combine multiple types of information.*"

"*We found that NMS resulted in a modest increase with training, but only for some tasks; furthermore, task complexity was not predictor of NMS emergence.* ***A causal relationship between success and dimensionality- and by extension, NMS-was not supported by our results****. One possible explanation is that the distinction in the degree of NMS may reflect the <u>qualitatively different ways of representing information with different dimensionality in working memory</u>*."

"*This suggests that loss of information encoded in a nonlinear manner may not be the primary factor of successful working memory-guided behavior*." We concluded that "*task success depends on task relevance of the represented information*", which "***seems to eliminate (by inhibiting) resources without benefit***."

The overall conclusion of this study is that the inherited cerebral features in monkeys prevent accessing much of the nonlinear information (the high dimensional information) that is stored in the unconscious. Their **brain inhibits the releasing of inefficient or impractical information that do not cooperate to immediate surviving**. Most of the nonlinear information, in the case of monkeys, is biologically impractical.

The hominins of the Last Clade, where Homo sapiens was the last representative, have differentially evolved out of this discussed stage.

We have today plenty of studies, which unveiled many elements of the unconscious mechanism in humans.

Joseph Jankovic MD, in Bradley and Darofff's Neurology in Clinical Practice, 2022, (*Intellectual and Memory Impairments'* paper) indicates the **existence of several types of implicit memory that are distinct from the waking memory** (conscious) state:

-procedural or implicit nondeclarative memory

-explicit memory or declarative memory that is divided in working (immediate) memory, episodic (short-term) memory, and semantic memory (long-term).

The class of memory for which subjects have no conscious awareness is the implicit or nondeclarative memory. The implicit memory does not involve hippocampal system. The procedural (implicit) memory involves "knowing how" rather than "knowing that".

Another type of nondeclarative memory is **classical conditioning** (an unconscious stimulus becomes associated with a reward or punishment). The last form of nondeclarative memory is called **priming**.

As one can observe, the **unconscious functioning relates to a type of implicit or nondeclarative memory** that is further subdivided into conditioning and priming.

Robin L. Carhart-Harris, Robert Leech, Peter J. Hellyer, Murray Shanahan, Amanda Fielding, Enzo Tagliazucchi, Dante R. Chialvo in a study *The entropic brain: a theory of conscious states informed by neuroimaging research with psychedelic drugs*, published in Front. Hum. Neurosci., 03 February 2014 have stated:

"*Entropy is a dimensionless quantity that is used for measuring uncertainty about the state of a system, but it can also imply physical quantities, where high entropy is synonymous with high disorder. Entropy is applied here in the context of states of consciousness and their associated neurodynamics, with a particular focus on the psychedelic state. The psychedelic state is considered an exemplar of primitive or primary state of consciousness that preceded the development of modern, adult, human, normal waking consciousness.*

Based on neuroimaging data with psilocybin, a classical psychedelic drug, it is argued that the defining feature of "primary states" is elevated entropy in certain aspects of brain function, such as the repertoire of functional connectivity motifs that form and fragment across time. There is a greater repertoire of connectivity motifs in the psychedelic state than in normal walking consciousness, this implies that primary states may exhibit "criticality," i.e., the property of being poised at a "critical" point in a transition zone between order and disorder where certain phenomena as power-law scaling appear. It is argued that this <u>*entropy suppression furnishes normal walking consciousness*</u>*. It is also proposed that* ***that entry into primary states depends on a collapse of the normally highly organized activity within the default-mode network (DMN) and a decoupling between the DMN and the medial temporal lobes (which are normally significantly coupled)***."

"*This article proposes that states such as the psychedelic state, REM sleep, the onset-phase of psychosis and the dreamy-state of temporal lobe epilepsy are examples of a regressive style of cognition that is qualitatively different to the normal waking consciousness of healthy adult humans.*"

"*Consistent with Karl Friston's free-energy principle (Friston, 2010), this article takes the view that* ***the mind has evolved*** *(via secondary consciousness upheld by the ego)* ***to process the environment as precisely as possible by finessing its representations of the world so that surprise and uncertainty (i.e., entropy) are minimized****.*

Later we finesse this basic model, arguing that ***secondary consciousness depends on*** *the human brain having developed/evolved a degree of sub-criticality in its functionality, i.e.,* ***an extended <u>ability to suppress entropy and thus organize and constrain cognition</u>****. It is argued that entropy-suppressing function of the human brain serves to promote realism, foresight, careful reflection, and an ability to recognize and overcome wishful and paranoid fantasies. Equally however, it could be seen as exerting a limiting or narrowing influence on consciousness.*"

"*These hypotheses are motivated by our finding that DMN-hippocampal coupling is decreased under psilocybin, and while DMN activity becomes dyssynchronous and therefore disorganized, the amplitude of BOLD (Blood Oxygen Level Dependent) signal fluctuations increases in the hippocampus.*"

Christer Johansson and Per Olav Folgero in a study *Is Reduced Visual Processing the Price of Language*, published in Brain Sciences on 12 June 2022 explain:

"*We suggest a later timeline for language capabilities in Homo sapiens, placing emergence of language over 200,000 years after the emergence of our species. <u>Detailed realistic cave paintings disappeared completely, and iconic/symbolic ones appeared at other sites. This may indicate a shift in perceptual abilities, away from an accurate perception of the present</u>. We suggest than an overwhelming flow of perceptual information,* ***<u>vision, in particular, was an obstacle to language</u>****, as is sometimes implied in autism with relative language impairment. Recent findings show brain lateralization for communication gestures in other primates without language, supporting the idea that a language-ready brain may be* *overwhelmed by raw perception, thus blocking overt language from evolving. We find support in converging evidence for a change in neural organization away from raw perception, thus pushing the emergence of language closer in time.*"

"*A recent article argued that superior visual perception was necessary for the creation of Paleolithic cave paintings because of the level of correct anatomical details and accurate depiction of high-speed leg position of animals in motion. The article uncovered and outlined current evidence for an association between visual thinkers (some diagnosed within the Autism Spectrum Disorder) and a relatively high percentage of archaic genes, of which some are associated with perception and cognition. However, the <u>linguistic abilities of visual thinkers may be impaired, which suggests a negative correlation between visual perception/memory and language</u>.*"

"*An ability to visualize a problem in great detail can often be very useful, for example, for innovators and scientists, where an extremely verbal, symbol thinker might not see the concrete details*."

"*<u>Compared to cave painters, modern humans have generally lost the extreme ability to remember visual details</u> and produce them artistically, but the natural ability is still present in visual thinkers and autistic savants. There is also contemporary evidence for pre-human superior abilities to remember symbolic/iconic sequences after extremely short exposure in chimpanzee.*"

"*<u>Access to detailed imagery makes generalization harder to accomplish</u> because of the high specificity of an image compared to general symbols. <u>In a detailed image, it is not obvious what is more important</u>, whereas verbal symbols may facilitate a focus on one feature at a time. Most people effortlessly focus on only one aspect of a word or construction at a time, forgetting about ambiguities that are not relevant in the context.*"

"*Hyperexcitation of the primary visual cortex, for example, through glutamate, will result in more acute vision, as in autistic savants and normal subjects stimulated by brain cranial magnetic stimulation (rTMS).*"

"*There is a grand scale interaction between auditory and visual information in the homologues of the areas we now consider most relevant for language and implies the overwhelming of these areas with interfering visual information, which may effectively prevent the emergence of language in other species. The Broca homologue of macaque is a meeting point of visual and auditory information.*" "*The precuneus has been identified as an area responding to both visual and auditory stimuli and possibly serving as audiovisual convergence area. What seems to happen is that the reliance on primary sensation has weakened, which has ameliorated the development of higher cognition.*"

"*The main theme of this article is to suggest that the full language capacity emerged much later than the emergence of Homo sapiens and that this coincides with the shrinking brain in the Upper Paleolithic (50,000-10,000 years ago) and the eventual disappearance of highly realistic cave paintings, which were replaced by iconic and symbolic cave art. The brain's capacity for processing complex recursive language could have been set free by lowering the intensity of neuronal information related to raw perception and raw visual perception in particular. The timing of a fully developed language apparatus coincides with the shrinking brain and a loss of superior processing of sensory information.*"

"*Raw perception and higher cognition are present also in other primates. The pre-cursors of language were and are still present in other animals. This capacity is prevented from emerging by the useful but overwhelming process of sensory information.*"

"*Human performance can be enhanced by altering the neuronal flow of information, as in transcranial stimulation studies. This could happen by psycho-active drugs, transcranial stimulation, or sensory deprivation.*"

"*We argue that perception, and vision, in particular, remained a strong priority until recently, which blocked full language abilities.*"

In a study by Souta Hidaka and Masakazu Ide (*Sound can suppress visual perception*), published in Scientific Reports (April 15, 2015), the authors demonstrate that "*white noise bursts presented through headphones degraded visual orientation discrimination performance. These results indicate that the perceptual suppression effect occur across auditory and visual modalities based on closed and direct neural interaction among those sensory inputs*."

Darinka Trubutschek, Sebastien Marti, Andres Ojeda, Jean-Remi King, Yuanyuan Mi, Mishka Tsodyks, and Stanislas Dehaene in their study *A theory of working memory without consciousness of sustained activity*, published on July 18, 2017, in *eLife* demonstrated that "*without the participant knowing, the brain appears to have stored the target location in working memory using parts of the brain near the back of the head that processes information. This non-conscious storage did not come with constant brain activity, but seemed to relay on other, "activity-silent" mechanisms that are hidden to standard recording techniques*."

"*Taken together, we found a clear distinction in the brain responses of seen and unseen trials. Converging evidence from our decoding analysis in the ERFs and alpha/beta band suggests that there was no apparent discriminative axis shared between the seen and the unseen correct trials. Similarly, the desynchronization in alpha/beta power characterizing the seen targets did not emerge on the unseen (correct) trials. These findings suggest that non-conscious working memory is a genuine phenomenon, distinct from conscious working memory*."

"*The majority of our evidence thus points against an interpretation, in which the unseen correct trials constituted either just a subset of seen trials or arose from the conscious maintenance of an early guess. It argues in favor of a differential state of non-conscious working memory with a distinct neural signature*."

"*It is, for example, possible to manipulate information unconsciously? It is not clear how stored representations could be transformed without being persistently activated and thus becoming conscious*."

Tetsuya Kageyama, Kelssy Hitomi dos Santos Kawata, Ryuta Kawashima, and Motoaki Sugiura in a study *Performance and Material-Dependent Holistic Representation of Unconscious Though: A Functional Magnetic Resonance Imaging Study*, published in Frontiers in Human

Neuroscience on December 6, 2019, have documented the "*Unconscious Thought* (UT)."

"***Unconscious thought (UT) refers to cognitive of effective decision-related processes that occur beyond conscious awareness, i.e., while people are consciously occupied in performing some other task.***"

"*Participants who are distracted show better decision-making performance compared with those who make an immediate decision (ID) and are allowed to think regarding the choices consciously.*"

"*The study findings revealed a greater involvement of both the right dorsolateral prefrontal cortex (DLPFC) and middle occipital gyrus during UT. The occipital gyrus is involved in visual processing. Research has indicated neural activation of multimodal cortices such as medial prefrontal cortices and the precuneus, relevant to the holistic representations. Studies reported neural activation of the ventromedial prefrontal cortex. The ventromedial prefrontal cortex plays a role in subjective and emotional values, whereas ventral Striatum and posterior cingulate cortex are involved in reward value.*

The cortical middleline structure, including the precuneus, is active during multimodal, which is consistent with the features of holistic presentation."

"*Neural activation of UT is also observed in other decision-making scenarios, such as lie detection, responses to requests, moral situations, and product satisfaction.*"

Seffelullah Kuldas, Hairul Nizam Ismail, Shahabuddin Hashim, and Zainudin Abu Bakar in a study *Unconscious learning process: mental integration of verbal and pictorial instructional materials*, published by Springerplus, on March 12, 2013, made a literature review on this subject.

"*This review suggests considering the role unconscious learning processes to enhance the understanding of how students form or activate mental associations between verbal and pictorial information.*"

"*Learning processes mainly refer to the interconnections between perception, memory, language, imagery, emotions, and motivation that allows students to mentally build connections between verbal and pictorial information patterns or between new and prior memories and integrate them with relevant knowledge structures in long-term memory. (Mayer and Moreno 2003).*"

"*Learning processes and outcomes can be conscious and unconscious. The unconscious processes range from registering information in the sensory memory to mentally forming associations within or between information patterns and activating associative memory networks, including individual expectations, beliefs, and desires (Kowalski and Western 2005). The unconscious can conduct the acquisition, access, and application of knowledge without deliberate and controlled attention (Ashby and Maddox 2005; Dienes and Perner 1999; Evans 2008). Thus, either conscious or unconscious learning is primarily a combination of mental processes, referred to as a knowledge acquisition process, bringing memories into the mind, forming associations, retaining, and using them (Mayer and Moreno 2003).*"

"*When humans consciously or unconsciously engage in learning activities, their potential emotions on learning process and outcomes are inevitable (Kowalski and Western 2005; Moreno 2010). The human brain operates automatically and rapidly to emotional stimuli, thereby generating unconscious responses (Bargh and Morsella 2008; Ohman 2002; Whalen et.al. 2004).*"

"*Appraisal processes are characteristically intuitive, immediate, and mostly outside conscious awareness (Arnold 1960), and are not typically the product of conscious reasoning process (Robinson and Clore 2001). Unconscious appraisals are mostly related to intrinsic properties of a stimulus, such as pleasantness or unpleasantness, and to individual needs, values, and goals (Scherer 1999). These appraisals generally lead to a motivational effect (Scherer 2009). Emotional processing acts as the on/off switch to motivation. (Moreno 2010, p.137). Emotions evoked during learning affect the way students learn (Howe 1998).*"

"However, the bulk of perceptual, cognitive, and emotional processes, and the organization of human memory are too complex to be delt with consciously; therefore, conscious learning is unlikely to occur all the time (Lewicki et.al. 1992). Conscious learning can be eased by unconscious learning processes that promptly and continually establish connections between the perceptual, cognitive, and emotional processes of learning, thereby facilitating the mental integration of verbal and pictorial instructional materials. However, mental representations may unconsciously evoke task-irrelevant thoughts and unnecessary load visual and verbal working memory capacity (Sweller et. Al. 1998)."

"Perceptual, cognitive, and emotional processes are carried out unconsciously at the onset, and afterwards, possibly reach conscious processes (Bargh and Morsella 2008). ***Unconscious processing appears to be structurally and functionally much more sophisticated than the conscious by accessing and influencing conscious processes*** *(Bargh and Morsella 2008)."*

"The common view on how human memory acquires information is that a limited amount of sensory information (visual, auditory, olfactory, tactile, kinesthetic, or gustative) is temporary held in short-term memory, herein organized by working memory, and, in part, permanently stored in long-term memory (Baddeley 1992). In general, relevant evidence indicates that cognitive unconscious processes can lead to the unconscious acquisition of knowledge and freely influence conscious learning process and outcomes, but later, this influence can be amenable to a conscious intervention (Evans et. Al. 2008)."

"***Hence, the conscious processing is unlikely to result in any kind of cognitive learning by obviating the need for the cooperation of unconscious processing***."

"Learning how to speak a native language, how to listen and interact with others, and how to use general problem-solving strategies are some examples of unconscious learning, which is particularly conceive to be easily, rapid, and unconsciously acquired biological primary knowledge (Sweller and Sweller 2006)."

"*One can unconsciously ask questions and seek answers without knowing the underlying factors of these behaviors (Chartrand and Bargh 1996; Greenwald and Banaji 1995; Jacoby et.al. 1992). In consequence, unconscious mental representations can be operative, either in the absence of or in the presence of conscious processes*."

"*Furthermore, **not just thoughts, but thinking itself can be unconscious** (Kihlstrom 2008), thereby causing more difficult for the conscious interference to affect the unconscious*."

All recent data from experiments suggest that unconscious information processing is implied in high-order operations associated with cognitive control, memory-guided behavior (during short- and long-term memory, but also with language computations. However, other research contests such data, indicating that the information about the world is reflected in brain activity patterns across a distributed set of brain regions. Some neurocognitive theories propose that unconscious processing reflects feed-forward processing only, while local recurrent connections are critical for bringing unconscious content into conscious awareness.

Deep learning is a type of machine learning and artificial intelligence that imitates the way our brain gains certain types of knowledge. Deep learning is a subset of machine learning that is a subset of artificial intelligence. Deep learning uses artificial neural networks to mimic the brain's learning process. Deep learning can be thought to automate predictive analytics.

The traditional machine learning algorithms are linear; deep learning algorithms are arranged within a hierarchy of increasing complexity and abstraction. Such hierarchy shows that each level of abstraction is created with knowledge that was gained from the preceding layer of the hierarchy.

Traditional machine learning is supervised. By contrast in deep learning the program builds the future set by itself without supervision. With each iteration, the predictive model becomes more complex and more accurate.

The inconvenience of deep learning concerns its requirement to access an immense amount of training data and processing power.

Neural networks involve a trial-and-error process. All such types of networks function by feeding data in and letting the model figure out for itself about the accuracy of the interpretation.

The biggest limitation of deep learning is that is based on the data on which they are trained. Once trained, deep learning models become inflexible and do not support multitasking. The reasoning is out of the question.

The data scientists prefer to use traditional machine learning over deep learning, because the result allows a superior interpretation, and the solutions make sense.

Chaos, nonlinear, and linear in human brain

Abstract

This paper aims to analyze the relationship that develops in the human brain between its main processing capabilities, which are nonlinear systems, chaotic systems, and linear systems.

As it is known, unconscious processing is completely nonlinear, and it represents more than 95 percent of brain activity. Even chaotic systems are an exception among the nonlinear ones, these chaotic systems control the emotional activity that is defined to be at least 90 percent of human thinking approaches.

In the end of this paper, I discuss one recent finding that make the latencies in brain processing to become a significant tool in linearize a part of the nonlinear processing. This linearization is called rationality, and it stands for as little as five percent of total brain activity. Even then, this linearity is our only awareness manifested in the works of consciousness.

Discussion

Nonlinear vs. linear

People usually make a confusion between chaos and nonlinear. As Harun Siljak said, "*if one looks long enough at any real system you consider linear, one will discover nonlinearities into it.*" Nonlinearities are variables. Probably, what Siljak had in mind is the hypothesis that data of linear lies in a smooth, low dimensional manifold, like the curvilinear. A Curvilinear relationship represents a relationship between two variables, whereas one variable increases, the other decreases. But it exists another type of curvilinear, whereas one variable increases, the other decreases up to a certain point, and after this, both variables increase together. There are many cases where a function looks curvilinear, but this function is not nonlinear. It remains linear.

Linear has a constant-rate of change, while nonlinear does not. Hence, the nonlinear is not a straight line. In the case of curvilinear the line has a U shape. **Linearization is a linear approximation of the nonlinear system, and it is valid only in a small region around an operating point. Away from this point, the linearization becomes poor**.

For example, gray matter volume development shows an inverted-U-shaped curvilinear trajectory, with a pre-adolescent increase followed by post-adolescent decease. White matter volume also shows a curvilinear trajectory, with a stable preadolescent volume followed by a postadolescent increase.

A linear system opts for components with minimal variability. Minimal variability is caused by low energy levels. Hence, every linear dynamic system still contains a variability that is small enough to be ignored for practical purposes. For this reason, a linear system evolution can be predicted.

Energy or entropy define a level of information. All biological organisms have a limited capacity for dealing with information. They are negentropic, meaning that they need to reduce the amount of environmental information to a minimum that assure the surviving.

Nonlinearity defines a dynamic system that is more complex than a linear one because it contains variables with higher energy (information) levels. The evolution of such a system becomes unpredictable because environmental information is more than the mental system capacity to process. Information is about options. Too many options confuse the simple surviving capability. Hence, the filtering of the biological system inhibits discarding all unnecessary options, or all options which do not fit the surviving goals.

Chaotic systems

It is known that "orderly" systems can transit into a state of chaos, and this process is generated by the evolution of complexity, or by the collapse of complex systems. The reverse also occurs. Because the **complexities are** dependent on certain levels of energy (entropy), this makes them **clusters of information**. Each component is a variable of information, and all components together make a cluster. Nevertheless, each component needs energy for its interactions. Therefore, biological systems are **open-systems**, meaning they absorb energy from the environment and expel the used products.

A dynamical system is chaotic if its future behavior is very sensitive to the initial conditions. However**, most nonlinear systems can never be chaotic**. **Chaos is an exception in the world of nonlinear dynamics**. On the other hand, nonlinear dynamics of variables dominate and rule the world.

The role of entropy and dissipation

The state of any dynamic system, being it nonlinear, linear, or chaotic, depends on its dealing with entropy. Entropy can generate a new type of order (in the form of a new complex system), but such order (stability) could be maintained only if the system gets rid of the excess energy (entropy) by dissipation. Dissipation helps the new complexity to remain stable at the energy level it had attained.

Most biological systems are dissipative, but not all of them.

On the other hand, maintaining an excess of entropy would lead to chaotic dynamics. It is known that entropy helps a system to transit from one level of complexity to another by supplementing more energy. But, if the energy is in excess, it breaks the stability of the dynamic system, supplying to it a multitude of distinct energy levels, which become the starting (initial) points (conditions) for the evolution of each new component; it makes the system's components to evolve on a variety of distinct or even divergent trajectories, which break the orderly behavior or coherence of the dynamic system. This generates a dynamic chaos that is still fundamentally deterministic. Theoretically, it could exist an "*infinite-order linear system*" that would be chaotic, too.

I would say that **all unconscious mind activity is nonlinear, while its chaotic behaviors are exceptions**.

"*Some nonlinear models of the cortex can be reduced to the linear case at the millimetric scale. But here, the EEG procedure exhibits nonlinear features which cannot be ignored. Its nonlinear character does not necessarily imply a chaotic spatial distribution. We cannot define the brain as chaotic, but we can test its nonlinear character. The initial sensibility to an infinitesimal disturbance defines the chaotic state (transitory in a nonstationary case). This nonlinear property means in general that the brain produces information. The information transfer between cortical structures requires nonlinear mechanism of synchronization and that this synchronization is a global phenomenon.*" (Zbigniew J. Kowalik, Andrzej Wrobel, and Andrew Ritz in Behavioral and Brain Sciences, 19 (2); 302-303,June 1996, DOI:10.1017/SO140525XOOO4276X).

"*A possible consequence of nonlinear information is to predispose the individual's behavior toward positive objects and away from negative ones when conscious mind is otherwise occupied (Van den Noort, 2004c). In other words, it might be an old evolutionary mechanism that humans share with other species.*"(Maurits van den Noort, Peggy Bosch, and Kenneth Hugdahl in Understanding the Unconscious Brain Evidence for Non-Linear Processing, published in https://www.researchgate.net/publication/253503721).

Antoine Del Cul, Sylvain Baillet, and Stanislas Dehaene published in October 2007 in PloS Biol 5(10);e260, have proposed **a threshold for conscious perception that manifests 100 ms after stimulus presentation**.

"*Electrophysiological recording in macaque monkeys during masking suggest a peak effect of recurrent interactions in area V1 around 100-140 ms after stimulus presentation, and intracranial human recordings suggest that category-specific ventral occipital-temporal cortices are already strongly activated by 150-200 ms although extension to more anterior regions and reverberation effects can extend as much longer (290-700 ms).*"

"*When this threshold is exceeded, even a brief external stimulation can simultaneously activate many distant areas and yield a long-lasting pattern of reverberating activity. Such a distributed pattern corresponds to a consciously reportable state.*" "*There is an indication that a major transition in processing occurs around SOA=50 ms, and this transition affects a broad range of cognitive processes*."

Response latency of the sensory-motor behavior.

The information I presented in the previous paragraphs refers to **the transition from nonlinear functions toward the linear functions of the consciousness**. It indicates **a sharp nonlinear transition between unconscious and conscious processing**. When the mentioned threshold is crossed, "*a reverberation takes place and stimulus information gains access through the workspace, allowing verbal report, voluntary manipulation, voluntary action and long-term memorization*." (Claire Sergent and Stanislas Dehaene).

It is experimentally proven that continuous reactivation (reverberation) reduces the modulation of the noise signal.

The fact, the peak effect of recurrent interactions is activated after a latency of 150-200 ms to 290-700 ms and indicates a "latency" in neural processing.

As the mentioned study also found**, it occurs a threshold within the "latency", where the signals crossing that border show a linear expression manifested as consciousness**. It results that **a longer "latency" contributes to increase the details of the linear manifestation**.

Here, one can compare the "latency" found in monkeys with that shown in humans: human latency increased significantly.

One can draw a parallel with similar findings between monkeys and humans regarding oculomotor system's "latencies". Again, human "latency" grew multifold compared to monkeys.

Latency produces linearization.

The mentioned study found that **an augmented "latency" increases the linear outcome. Hence, the increased "latency" in humans resulted in an extended consciousness**, bringing in more details.

Could we make a connection between neural "latencies" and oculomotor "latencies"?

Several theories view recurrent interactions between distant brain areas as a necessary condition for conscious perception. The recurrent connections allow long-distance communication and auto-amplification of the activation.

However, the **chaotic behavior of the human mind is significant and could affect us during a lifetime**.

Chaos of the mind

Dr. Peter Michaelson in *Why We Suffer.Com, Transformative Insights from Deep Psychology* on an article published December 20, 2013, provide a complex picture of chaotic processes in our brain.

"***Each childhood's influences on our adult experiences have parallels to the scientific concept of Chaos Theory***."

"*The unconscious mind of adults is buffered by gale-forces winds of emotional chaos that originated as an infantile effect decades earlier. Emotional associations from our distant past now buffer our life in incredible, mysterious, spectacular, and frequently painful and self-defeating ways. Emotions percolate and circulate in our unconscious mind with some degree of chaos*."

"*Our brain struggles mightily and often unsuccessfully to limit the effects of irrationality. Often we try to apply common sense and reason to moderate unpleasant emotions or to curb self-defeating impulses*."

"*Children's first signs of self-awareness show up after twelve months. They detail the emergence of growing rationality in the second year and beyond. The baby is entrained in an instinctive irrational consciousness and intense entanglement in a false reality driven by acute self-centeredness, primitive wishes, libidinal instincts, inner aggression and fear, and self-defeating compromises. This infantile consciousness continues in various degrees over a lifetime*."

"*According to psychoanalysts, babies experience the world through megalomania, or profound self-centeredness, along with a sense of omnipotence or magic power. In this primitive consciousness, nothing exists except what the child experiences directly. Whatever is given to the child by others is experienced by the child as an act of his or her magic power. The illusion of having such magic powers still resonates very strongly in adolescents and teenagers*."

"*We become hooked on the feeling or attached to it. It can even be described as an emotional addiction. Whatever remains unresolved from our past becomes an emotional attachment. We're creatures who seek consciously or unconsciously to experience situations in ways that replay our unresolved emotions, whatever positive or negative.*

Evidence supporting a chaos theory of the human mind is overwhelming."

Non-linear thinking

Beatrice Barbazzeni in her article *A Linear Brain in a Non-linear World: How to reverse your Thinking with Neuroscience to Challenge, Experiment, and Explore*, published in Exoinsight on December 28, 2021, explains:

"***Non-linear thinking** is not sequential, and it **refers to the ability to create connections between concepts/ideas, even when they are not related**. Nevertheless, whether linear thinkers lack in understanding and connecting facts that are not sequential, non-linear thinkers due to their attitude in abstracting facts they might not perceive things that have a definite cause without understanding their effects*."

Studies on brain structure indicate that **human cortex matures mostly in nonlinear manner, even then, some structures mature in linear manner**. Such maturation occurs progressively with age, but it differs from one individual to another, and the maturation timing varies from one structure to another. Hence, the brain displays multiple types of nonlinear and linear functional patterns.

Ashkan Faghiri, Julia M. Stephen, Yu-Ping Wang, Tony W. Wilson, and Vince D. Calhoun in their study Brain Development Includes Linear and Multiple Nonlinear Trajectories: A Cross-Sectional Resting-State

Functional Magnetic Resonance Imagining Study (published in December 2019 in Brain Connectivity 9(10):777-7888) explain: "*Drastic changes in brain structure and function occur during the first two decades of human brain development. Most studies have focused on changes in the activation of certain regions as children develop.*"

"*It is generally believed that the overall organization of human brain is established at birth, but maturational changes throughout childhood and well into adulthood impact efficiency and specialization of the human neural system. There is a distinct change in the connectivity of brain networks from local to distal as children mature.*

Nonlinear developmental trajectories have been studied extensively in structural studies. For example, there is ***a nonlinear relationship between age and activation in regions involved in social cognitive processing***." *Also,* "*connectome distinctiveness*" *has a nonlinear increase with age.*"

During aging, various structures change their connectedness, some become stronger, some weaker, some become nonlinear, a few turn linear.

Nonlinearity of the mind and the pool effect

Fei He and Yuan Yang in their study *Nonlinear system identification of neural systems from neurophysiological signals*, published in Neuroscience, 2021 Mar 15: 458; 213-228, explain more issues about human brain nonlinearity.

"*Neuronal information processing is complex at different levels, from the microscopic pre- and post-synaptic cellular interactions to the macroscopic interactions between large populations of neurons in, for instance, sensory processing and motor response (Stanley, 2005). The behavior of a single neuron is highly nonlinear, showing a step-like none-or-all firing response (Koch and Segev, 2000), while the behavior of neurons in a population could be relatively similar. The nonlinear response of each individual neuron can be smoothed out by the distribution of membrane threshold across the population, known as the pool effect (Rosenbaum et al., 2010).* "

The pool effect generates a type of functional linearity. "*In a neuronal population,* ***the pool effect can reduce the nonlinearity generated from each individual neuron, by smoothing the neuronal dynamics from a scale of milliseconds (spikes) to 10 milliseconds (local field potentials0 or to 100 milliseconds (large-scale neurophysiological activities or signals such as EEG (Ros et al., 2014). Such effects have been previously demonstrated, like when the motor command can be transmitted linearly*** *via the mono-synaptic corticospinal tract* ***when more than five mononeurons are activated*** *(Negro and Farina, 2011). However, a small amount of nonlinearity may still be present (Yang et al., 2016).*"

"*Linear neural connectivity can be assessed by determining the signal correlation or causality between the recorded neural signals.*" "*Nonlinear behaviors in neural systems are thought to be associated with various neural functions, including neuronal encoding, neuronal processing of synaptic inputs, communication between different neuronal populations, and functional integration.*"

"*Recent studies indicate that linear methods can only capture a small amount of neural activities and functional relationships, and therefore cannot describe neural behaviors in a precise or complete way. (Vlaar et at., 2016, 2017).*"

Here would be interesting to draw **two multidisciplinary parallels**.

The pool effect in swarming

The first one refers to the swarming phenomenon, where several individuals develop cooperative behavior. A theory suggests that when individuals emit specific signals (like the light produced within the emission of biophotons, or the acoustic signals, or even chemical reactions), such signals activate the field of the weak forces called van der Waals forces, producing a cooperative interaction between the emitters. It suggests the ability to selectively activate molecular vibrations during signal absorption. The produced excitation grows when the number of elements (molecules and/or organisms) increases.

The van der Waals effect appears only when a signal occurs, and it ceases when the signal disappears. During the van der Waals effect develops a synchronous control manifested over the "pool" of affected elements.

In the previous paragraphs, I showed that, at neuronal level, **the pool effect generates linearity in command** transmission (it is like lower-dimensional latent manifold) among the cells or neurons. Now, I can assume that a similar effect of command transmission occurs during swarming.

Linearization is produced by the latency of the oculomotor system.

The second parallel I would like to make refers to oculomotor system latency. The previously mentioned quotation (Fei He and Yuan Yang, 2021) indicates that the "pool effect can reduce the nonlinearity by smoothening from a scale of milliseconds (spikes) to 10 milliseconds."

Hence, when the scale increases the linearity increases, too. The research showed that oculomotor latency in humans increased significantly, when we compare it with latency in monkeys. Thus, I would suggest that **the increase in human oculomotor latency had produced more linear details in perceptual processing, extending the consciousness, too**.

As mentioned before in this paper, neural activity also manifests latencies, and within them there is a threshold, and both contribute to linear outcomes.

However, the mentioned "latencies" generate a part of the linearity that creates the consciousness. But such linearity remains 5% or even less from total activity of the human brain.

Mario Natarelli, in an article in *Insider,* published in April 2020, explains that "*We now know about how the human brain processes and triggers our behaviors than ever before.*"

"*We now know that up to 90 percent of the decisions we make are based on emotion. Almost every decision we make is based on emotion, not rational thought, and measured consideration. This suggests that traditional models, constructs, and methodologies that we have used for decades to drive our marketing and communications efforts outweigh the importance of rational thinking, rendering them outdated and faulty*."

"***The idea of humans as rational actors who make decisions to buy and use products and services in purely rational thinking is clearly flawed***."

Other research found that **all animals follow the same "emotional-decision-based" process**.

Pessoa (2008) suggests that the core region of emotion representation (like the amygdala) **integrates emotion and cognitive information**. Hence, **emotion representation could not be separated from cognitive representation** (Schaefer et al., 2006). **The emotions are integrated in the behavior and thought**.

Our civilization originates into a hierarchical regression of vertical complexity

Abstract

The aim of this paper is to analyze the evolution of human species through the perspective of biological evolution. Biological studies indicate that biological systems can hierarchically regress of vertical complexity. The outcome of such regression usually leads to the development of horizontal complexities.

Within human species, the development of a horizontal complexity generated linearity, affecting the evolution of the consciousness. Horizontal complexity allows linear composition of components, and in part was responsible for assembling a coherent language. The same compositional capacity can assemble technologies. Such a compositional/assembling feature generated our civilization.

The paper analyzes the characteristics of vertical and horizontal complexities and the process of hierarchical regression.

This subject is of relevance for the evolution of human society and its future.

Discussion

The studies in biology indicate that the size of each level of biological complexity increases as complexity increases. When the complexity ascends hierarchically, the complexity size increases by many orders of magnitude compared to the preceding level. Each level of structural complexity has its own unimodal size distribution.

Noel A. Heim, Jonathan L. Payne, Seth Finnegan, Matthew L. Knope, Michael Kowalewski. S. Kathleen Lyons, Daniel W. McShea, Philip M.

Novack-Gottshall, Felisa A. Smith and Steve C. Wang published a study *Hierarchical complexity and the size limits of life,* in Proceedings of The Royal Society B on 21 June 2017.

The authors said: "*It suggests that organisms at each level of complexity must be substantially larger than those of the preceding level, largely because individuals are composed of multiple components from lower level. The expectation is that the variance in size within each megatrajectory increases over geological time. The vertical complexity was a prerequisite for evolving organisms.*"

However, this study refers to the increases in the body size of biological organisms. But, in the meantime, the body size of the organism is somehow equivalent to the size of the complexity itself.

"*Our results are most consistent with the hypothesis that minimum size constraints arise from physical factors, while* ***constraints on maximum size arise from Physiological factors****. Physiological factors are such as thermoregulation and ability to acquire enough food or energy.*"

"*Effect-macroevolution is postulated to occur when the trait undergoing evolution is a property of the organism rather than emergent property of a higher-level taxon, and* ***the trait is associated with differential macroevolutionary dynamics at those higher levels*** *(e.g., extinction rate covaries with body size).* ***They (Maurer et al.) found size-related biases in origination and extinction rates****.*"

"*Diffusion-like processes operation within each hierarchical level do not imply that there is no selection. At the higher level of vertical complexity, the multitude of selective processes operating at lower levels may produce emergent patterns that resemble diffusion.* ***Diffusion evolution is the primary mode operating within complexity levels****.*"

"*Viruses are unusual in that they probably arose by decrease in vertical complexity from cellular ancestors. Partial decrease in vertical complexity is not uncommon. In fact, decreases of this sort were slightly more common than partial increases.* ***Larger downward transitions-full drops from one vertical level to the next one down-appear to have***

occurred multiple times. The transition from bacteria to viruses are examples of this."

"*A derived state of reduced vertical complexity does not carry the ancestral size constraints of the previously occupied vertical level. In other words, the size of a virus that evolved from a cellular organism is constrained by the complexity associated with being a virus and not by the shaped evolutionary history with much larger cellular organisms.*"

"**Evolutionary innovations associated with new levels of complexity, therefore, appear to be fundamentally different in nature from those that arise within complexity levels.**"

As these authors said, **downward transitions of vertical complexity have been usual occurrence in biological evolution**.

A change is said to occur reversibility. Reversing every step of an incrementally evolving process, in some cases, it would restore the system and the surroundings to their previous state. Nevertheless, in some other cases the system can be restored to the previous state, but the surroundings remain forever changed. Even in some other cases the processes are not reversible.

This proven idea is what I made to be the main topic of this article. It shows that **biological complexities could regress, mostly driven by changes in the ambient thermoregulations or/and diminishes in the availability of energy** (here food is seen as a source of energy).

Shaul Hochstein and Merav Ahissar (with Hebrew University, Israel) in their study *View from the Top: Hierarchies and Reversed Hierarchies in the Visual System*, published in Neuron, Vol. 36, 791-804, December 6, 2002, explain:

"*Our initial conscious percept-vision at a glance-matches a high-level, generalized, categorical scene interpretation, identifying "forest before trees." For later vision with scrutiny, reverse hierarchy routines focus attention to specific, active, low-level units, incorporating into conscious perception detailed information available there.* ***Reverse Hierarchy***

***Theory** dissociates between early explicit perception and implicit low-level vision.*"

As the above example indicates, it occurs a reversing path of entropy because the conscious (the explicate) is negentropic(it retards entropy), while the unconscious is an implicate (it operates with higher entropy).

From Paleolithic down to modern eras, the consciousness (the explicate) vastly evolved following a significant reduction in entropy, and it helped surface more details of the perceived reality.

What happened to human mentally while vertical hierarchic complexity regressed? As it appears, the horizontal complexity linearly increased, while the vertical complexity decreased. But here, the horizontal complexity is embedded in vertical complexity, and horizontal evolution depends on some of the vertical constraints, in the sense that it cannot exceed the size of the vertical complexity. In other words, horizontal complexity evolution is limited by the size of the vertical complexity.

I would like to analyze several examples of regressed complexity, which occurred in human history.

It is documented that 6,000 years ago, Amazonians practiced agriculture (farming). During the settling of the colonists on the eastern edge of Amazonia (starting 500 years ago), local Amazonians experiences a spread of European and African diseases, which were fatal for Indigenous population.

The Neolithic Amazonian survivors, who were farmers, were forced to move westward and deep inland. As a result of this forced migration associated with large population losses, the Neolithic Amazonians regressed to a pre-farming stage that brought them back to a Paleolithic society.

The Indigenous people of New Guinea had a Paleolithic culture until 7,000 to 4,000 BCE, when Asian visitors thought them agricultural practices. Around 5,000-4,500 BCE, the Papuan population developed farming, crossing into the Neolithic stage.

It is very debated the issue of farming in Australia. Several Australian specialists support the idea that the lands of Australia, at the time when the first European colonists arrived on this continent, showed traces of a previous agricultural practice.

However, the aboriginals of Australia were in a Paleolithic stage at the European arrival. There is to hypothesize that the Papuan farming spread into Australia for several hundred or maybe thousands of years through the land-bridge over Torres Strait until such land bridge collapsed due to sea flooding.

There is possible that the Aboriginal Australians crossed into Neolithic in the Cape York Peninsula, and this phenomenon eventually spread further south.

As it seems, it was a short-live phenomenon because it developed as a transition stage, where mixing with previous foraging and nomadic life persisted. Archaeologists found, but only in the north of the Cape York Peninsula, some evidence of a banana farm.

Such transitional Neolithic conditions gradually vanished all over Australia, and the aboriginals regressed one step back into the Paleolithic stage. However, as it appears, some agricultural or Neolithic traditions persisted over the course of time.

(I am not familiar with Australia's conditions; hence, I cannot document this idea in detail.)

As biologists investigated, the regress of complexity was quite habitual in the biological world. It brings a surprising fact for the Science of Complexity. This science indicates the mechanisms that make the complexities develop and collapse. The issue of regression is very little debated.

As it appears, the regression is only one step backward or a limited collapse process, or it is a micro-complexity collapse.

I can postulate that during the migration out of Africa, locally occurred many regressions and even full collapses. Such events provide archaeologists with misleading evidence about human evolution.

As we know, during historic times, many other regressions and collapses have occurred.

For example, the Mesopotamian Civilization, after almost 3,500 years was conquest by Persia in 539 BCE. Persia was destroyed by Macedonian conquest (331 BCE) but it revived during Parthian and Sasanian empires to finally collapse during Arab conquest (651 AD).

Romans took control of most of the Hellenistic Civilization in 116 AD. The Western Roman Empire was defeated by various migrant people, while Eastern Roman Empire (Byzantium) ceased to exist in 1454, being defeated by the Ottomans.

The Mongolian Empire and conquest of 13th Century AD was gradually defeated and eliminated by Russians and absorbed by the Chinese, ceasing to exist in the 17th Century AD.

Arab Civilization was defeated by the Ottoman conquest of the 7th Century AD. Since 1881, the European conquest defeated the Ottomans in North Africa.

European conquest of the colonial era defeated all populations of Americas, Africa, most of Asia, Australia, and Oceania.

Many of the defeated empires and the civilization they created have been destroyed permanently, like Grecian, Phoenician, Macedonian, Roman, Mesopotamian, Persian, Mongolian, and Ottoman empires.

These examples show partial and permanent collapses of the vertical hierarchies represented by those empires and the cultures they created.

Western European colonialism was generated by the "marketplace" initiated in North Italy beginning with the 13th Century AD. It financed world exploration for the means of new resources and markets. It upgraded the art to the level of an asset, allowing the "marketplace" to sponsor and finance the Renaissance.

"The marketplace" is maybe the most significant event of human history after the advent of agriculture and metal manufacturing age.

The climate played a major role in the collapse of the Indus Valley Civilization, the Mesopotamian Civilization, and the Wet Saharan multi-polar Civilization. In more recent times, the Medieval Warm era, and the Little Ice Age in Western Europe had significant effects on the local cultures. Climate fluctuations affected the Silk Road operation.

In general, climate fluctuations vastly affected the African populations in their continental dispersal and connectivity; such fluctuations were associated with similar climate events in the Middle East.

The most major climatic factor in human evolution probably was the change from the glaciation stage and the maximum cold of the LGM to the Interglacial era and the Holocene.

* * *

I can postulate that the **mentioned climate fluctuations caused numerous regression and collapses of the local cultures, influencing human evolution**.

The **Anthropocene** has its own role in influencing the complexity of human society and its evolution. It started with agricultural practice and metal processing, evolved to industrial era, industrial revolutions, digitalization, automatization, computerization, and AI.

We are embedded into everything our civilization generates, and it is very hard for one to distinguish current regressions and collapses. For sure, the **wellbeing** (1 billion people live in slims, 300 million people were forced to migrate in 2022, or 4 percent of the world population because of local degradation of the basic living, modern slavery, and human trafficking accounts for 50 million), the **health degradation** (one in three world's inhabitants suffer of chronic multiple diseases) are fundamental markers of current human achievement and are a subject of intense debates.

The vast multiplication of human emotions serves as a perfect stimulus for vicious consumerism.

Human consciousness continues to evolve exponentially, and it represents a significant increase in the retarded entropy (negentropy).

As it appears, most of such increase is generated by the multiplication of human emotional panel. Here, rationality (as an asset of Frefrontoparietal Synthesis) serves only as the guidance for all combinatory and multiplication processes.

* * *

Probably, the most significant collapse of mental complexity occurred during our prehistory, when many species of hominins have disappeared due to geomagnetic events, which were associated with ionizing processes. The ionizing radiation may increase the entropy in the brain beyond the capacity to dissipate it.

The aspect of biopositive and bionegative effects of the ionizing radiation is highly debated, and it seems to vary from one species to another, while variations were found between the individuals of the same species.

One aspect of higher atmospheric concentration of C14 isotopes is related to the stimulation of neurogenesis. Here, Homo sapiens, or a large part of this species, in the last 100,000 years, reacted to excess neurogenesis by developing new mental structures, like the frontoparietal and parietal regions development simultaneously with an enlargement of the precuneus.

Beyond the stimulation of neurogenesis, the oxidizing effect of the ionizing radiation generates free-radicals, while the biological organisms could deexcite them before any harmful effects would occur.

The deexcitation of free-radicals produces biophotons. Also, in certain conditions, the free-radicals seem to have an essential role in biological functioning.

However, as the research has revealed, higher entropy is related to higher intelligence, and it occurs during meditation practice. It is hypothesized that higher entropy is useful, when it comes in limited pulse, which can be conveniently dissipated. This indicates that certain pulses of entropy are beneficial to biological systems and to brain activity. Eventually, such pulses, being followed by minute dissipation, drive the biological system to a novel energy/information level that becomes stabilized by the dissipation. Undoubtedly, the dissipation stimulates free-organization, and like the meditation shows, the meditation practitioner develops a better attention.

The emission of biophotons, because of deexcitation of free-radicals, is another subject of debate, because the biophotons can allegedly serve as new neural pathways for faster inter-communication between different regions and neural networks of the brain.

The safe absorption depends on the amount of ionizing radiation produced (like the low-doses have a distinct effect compared to high-doses) and on each system capacity to deexcite the free-radicals.

As it appears, Homo neanderthalensis and the rest of the contemporary hominins 9like the Denisovans) could not achieve such a redirection of the excessive neurogenesis, and thus, their brains reacted by causing mental diseases, which were responsible for a rapid extinction.

* * *

Entropy is information; thus, the amount of information processed becomes retarded, too. Nevertheless, higher entropy helps generate self-organizing processes, like self-organization proper and self-organizing complex systems. But here, the bulk of entropy being accessed by the system is more than the system needs; hence, the excess entropy (energy/information) is driven away by dissipation. Biological systems are a subset of dissipative structures.

Dissipation helps the system survive far from equilibrium by favoring the interactions between the systems' components.

In short, dissipation is the generator of self-organization and of self-organizing complexities far from equilibrium.

But the right pulse of entropy can help a system to be reorganized on a new higher energetic/informational level.

In general, **a regression of the vertical hierarchical complexity leads to continually extending of horizontal complexity,** as the path of regression accelerates. During the transfer from vertical to horizontal, this process indicates a regression of the information (energy, too) content caused by dimensional reduction.

The horizontal hierarchy incessantly grows in linear details, but the needed overseeing of a nonlinear capability is mostly lost. Like a traditional saying: one cannot see the forest because of the trees.

* * *

The activity of the brain is formed by wave vibrations (electrical) and their oscillatory rhythms. Their coupling, decoupling, and interference help produce self-organization, self-organizing complexities, and dissipation.

Hence, during regression, or hierarchical transfer to a lower level, the vibrations with lower dimensions must self-organize to adapt to local conditions. In the meantime, the self-organization of low dimensional systems generates better organization.

However, at higher hierarchical level, there is higher entropy, which drives those system into more advanced equilibrium states, and where the interactions are minimal or non-existent.

The regression favors the transition to lesser equilibrium states. Here, dissipation and self-organization make the lesser equilibrium, and far from equilibrium systems persist by increasing the pattern of interaction.

In other words, the horizontal hierarchy is the place where the brain's electrical activity is characterized by rhythmic waves, called vibrations or neural oscillations, which interact by coupling, decoupling, and interference; these are combinatory processes, and define, in part, the brain activity generating consciousness.

I would say, the mentioned vibration combinatory processes become synonymous to mental analysis and synthesis. As it appears, **the enlargement of precuneus, since 70,000 years ago, seems to play a large but novel role in the new frontoparietal synthesis capacity developed in the modern human brain around 30,000 years ago**.

The **frontoparietal synthesis helped generate mental combinatory activity, where analysis and synthesis characterize the rationality**.

However, as recent neuroscience studies found, rationality still represents less than five percent of the human brain activity. The rest remains emotional processing.

Emotions result from natural selection as a survival adaptation to various changes in the environment. They also regulate or guide the relationship between individuals.

The emotions are stored in the implicate memory and manifest in the explicate manner as conscious activity. The implicate defines the unconscious and nonlinear activity of a higher or transitory hierarchy that is distinct from consciousness.

The unconscious setting of the emotions suggests that they originate into a physical domain with higher entropy, minimal or no interaction, and more (thermal) equilibrium.

Gibbs's Free Energy principle says: The system is more stable when the surrounding entropy is higher. The stable state is always the one which has the largest total entropy for the material and the surroundings. Or, one could say that **a system is stable when internal energy is at a minimum and its entropy is at a maximum**.

I can postulate that the emotions have a minimum internal energy and are stored in a medium with high entropy, like the case of a high vertical hierarchy. Hence, as it seems, the fundamental features, controlling the survival, must be safely stored in a medium where a more stable equilibrium can exist. But such stability is voided of interactions, implying the unaltered conservation of the stored information.

In their transition toward the explicate, the nonlinear emotions must adapt to the conscious linearity, where combinatory processes are prevalent.

As an explicate outcome, the emotions become divided, combined, and recombined. This process significantly but artificially multiplies the original basic emotions. It is like subdividing them into various colors, and each color into many shades. Such a multiplication had a practical role because it allowed the **transition from a course-grain to a fine-grain adaptation**.

This fine-tunned adaptation eventually allowed social self-domestication, while augmenting the inter-individual interactions.

In all this, **rationality, as analysis/synthesis, remains only a guidance that oversees the novel combinatory processes**.

I would postulate that **"rationality" is not a component of the consciousness but a linear tool that helps generate all combinatory features**, and in this way, it actively participates in the build-up of the modern secondary consciousness.

By contrast, the "primary consciousness" would have a raw type of emotions and very little interactions. As the neuroscience experiments demonstrated, its entropy is significantly higher that implies a more stable equilibrium. The same features were found for the states experienced during meditation and/or under administration of experimental drugs.

I suggest that consciousness evolution in the modern human brain (named secondary consciousness) was generated by such exponential compositional multiplication within the emotional realm.

* * *

As stated before, the horizontal complexity is embedded into and limited by the size of the vertical one. It means, our conscious development has a limit, but which is one step smaller than the limit of the vertical one. There is not a danger that we will reach that limit anytime soon.

In the meantime, the conscious development, as a complexity, is independent of the complexity of unconscious host.

However, our thinking remains over 95 percent emotional and provided by the unconscious nonlinear host. Rational linear thinking provided by consciousness is a very small percentile, and it does not seem to significantly evolve. Emotional thinking is naturally efficient, while the rational one is naturally inefficient. Here, natural selection prevails because the emotions originate in natural selection and rationality is an artificial mind construct.

Rationality arisen as a servant of the moment, and it originated as a derivative of the language's practice. It may be important to indicate **that rationality incipiently appeared, along a coherent language and stimulated by it**, at the time the Last Glacial Maximum begun to manifest (some 25,000 years ago). It paralleled the first tentative toward agriculture and semi-permanent human settlements (23,000 years ago on the lakeshores of the Sea of Galilee). It begun to change the life of the foragers, marking the incipient development of that regression that took the form of the horizontal hierarchy.

We made rationality the philosophy of life and a fundamental asset of science, driving both within a strict linearity. **Rationality, as a combinatory approach, helped improve momentary survivability.**

Time and again, rationality (manifested as synthesis and analysis) arose within the horizontal development; it started with adding and combining nodes to the linear network (synthesis) and evolved to search into each node competency (analysis). Such addition delegates or combines work from one node to many. In the process, it helps to add skills to each node by improving its performance (competency).

All these maneuvers improved the chance of immediate survivability by generating new tools and assets intended to protect against hard to adjust environmental changes or conditions.

We are surrounded by a nonlinear reality, where is the game played by the complexities. Everything evolves from simple to complex and collapses back to simple just to start over again. **All collapses are caused by the internal relationships among the parts that run into an energy deficiency**.

Our horizontal conscient conscience is linear, but these linear elements (thoughts) output generates complexities. Our emotional nonlinear thinking, being placed on the edge between horizontal and vertical hierarchies, can provide nonlinear solutions (intuitive outputs) to many of such complex outcomes.

Rational thinking, as the main product of consciousness, is an energy eater. According to Yale University (June 19, 2009) a normal human brain consumes 50 per cent more energy than a heavily anesthetized brain with knocked out consciousness. By comparison, the studies on various mammals found that the average energy consumption for conscious processing could be around 6 per cent. (I did not find in the literature any paper to interpret such a difference.)

In another paper of mine, I discussed the latency of neural processing in humans compared to chimpanzees. Human's latency is at least 200-250ms, while in chimpanzee is 40ms. In the case of a speaking/listening individual this latency could increase up to 2,500 ms.

Hence, as much as the conscious processing increases (like by speaking/listening) the latency increases, too. The latency is the gap between the time a signal is perceived and the time the brain consciously processes it. This gap is a space of missing information, and it is filled by the brain mechanism with "estimates" about the missing information.

The latency, among other things, increases the number of details of the perceived reality by adding those "estimates". The mentioned "estimates" are provided by the implicate memory that is on the edge between unconscious and conscious. This is one way in which the unconscious data is brought into the conscious.

As it appears, latency **is one significant factor that decreases the consciousness's energy consumption and reduces entropy**.

A part of the conscious energy is used by rational maneuvering. The conscious brain consumes almost 50 percent of brain energy, but it processes only 2-5 percent of the total perceived information; **it shows how inefficient and energy consuming is the transition from nonlinear into linear**.

Such high energy consumption may have provoked the mentioned complexity regression. **Once the regression is completed, the energy consumption becomes minimal**. As the transition increases, the path of regression accelerates.

Let's see what a **horizontal complexity** (organization) is. It is called a flat structure that allows direct lines of communication between its linear components. All components are on the same scale with very few hierarchical levels. It forces the components to work together because the interconnection is fast, it allows composition, iteration, and all other but simple linear approaches. The drawbacks are reduced productivity, no expert (emergent) help, undefined expectations. It consumes a minimal level of energy.

Vertical complexity has the highest level of resolution because it uses the emergent result as the solution. It is hierarchically staked on many self-similar levels. For a multi-dimensional object, each dimension is contained by one level or scale. The physical properties at each level (scale) are distinct because the energy grows hierarchically upward.

Communication between levels is difficult and implies transition and energy consumption.

However, vertical complexity is about leadership, where the leader bears the fruits of emergent behavior that translates into long-run capabilities, like wisdom, creativity, adaptation to system changes, intervening constraints, and environmental variations. It generates inner complexities capable of resolute the outer complexities.

Emotions represent the natural selection of survivable instincts and drive our relationships with others and with the ambient. Emotions are processed by the brain nonlinearity within the vertical hierarchy. Emotions mostly produce a type of leadership that is intended to protect against dangerous outcomes and cause adaptability. Not all emotions are protective, resulting from errors in mental processing. "Rationality" tends to exercise a linear control on such errors that result in detrimental emotions.

In sum, even when emotional thinking dominates our mental processes, its conscious translation is the most energy consuming process in the brain. During their recent evolution, humans developed a capacity to control emotions, and this capacity represents the mentioned translation. Such a control involves “rationality”. If emotional processing represents 95 per cent of the entire brain activity, then, the “control” of this processing occupies less than 5 percent, but it is very energetic indeed.

In the meantime, our exclusive linear approaches are materialized within the ambient as an immense number of complexities, which naturally are nonlinear and have a high level of variability and unpredictability.

Our only resolution to such ambiently formed complexities can be provided by the mentally driven vertical complexity. It means that we must exploit in a novel way the nonlinear aspect of emotional thinking.

The computing machines cannot help to exploit this novel direction because they are based on rationality (linearity) not on emotionality (nonlinear rooted).

It may be a chance that quantum computing would unveil many of the quantum mechanism characteristics, which regulates at the fundamental level all hierarchic complexities, including our horizontal one. At least, **quantum computing** may become fairly translated within the boundaries of our linear understanding. Could occur a quantum computing assisted thinking, like a quantum-rational-exploit?

And then, we may hope to exploit in a novel but practical manner the hierarchically vertical and nonlinear roots of our horizontal hierarchy, while conserving most of the evolutionary adaptations of the current linear thinking.

As far as one can postulate, thinking in a natural but nonlinear manner would imply switching to a type of a fundamentally distinct brain. Probably, if certain conditions would occur, allowing a biologically feasible change of natural origin, such a transition could take a very long time, like hundreds to thousands of years.

* * *

In sum to this chapter, I would like to conclude that, in my opinion, the hierarchical reduction of mental complexity in humans can be defined as a cause of the exceptional development of precuneus in the last 70,000 years.

As Andrea E. Cavanna, and Michael R. Timble explained in their review article *The precuneus: a review of its functional anatomy and behavioral correlates*, published by Oxford University Press (January 6, 2006), precuneus is responsible, single or in association with its neighboring neural structures, for the most important features showed by the modern brain.

In short, the article shows that the experiments proved that "*precuneus is involved in a wide spectrum of highly integrated tasks, including visuo-spatial imagery, episodic memory retrieval and self-processing operations, and also it has the highest resting metabolic rates (35% of the whole brain energy consumed during this task). It has been recently proposed that precuneus is involved in the interwoven network of the neural correlates of self-consciousness and engaged in self-related mental representations.* ***The richly connected multimodal associative area to which precuneus belongs may be part of the neural network subserving self-awareness and conscious experience***."

Therefore, precuneus exceptional while unique evolution in humans starting 70,000 years ago led to abstraction, imagination, a capacity to combine items by coordinating the use of hand, or performing compositions only at a mental level, the generation of the self-awareness, contributed to extend the consciousness, and allowed an increased level of socialization, helping to develop an increased level of self-domestication.

All these features represent a mental extension deep into the linear three-dimensional space, which started to develop in the Homo sapiens brain only 70,000 years ago, and in association with other cortex developments led to generate a language-ready brain.

The **precuneus is a significant, even might be the essential part of the novel neural mechanism that generated linearity of thinking in humans**. Various functional latencies, occurring due to neuronal complexification and which appeared contemporary to precuneus recent development, have contributed to serve linearized and nonlinearized outputs to the cortical mechanisms (that is known as the pooling effect).

<u>Precuneus, as a neural hub, had been able to maneuver various latencies, which, as said, have dual nonlinear and linear outputs</u>.

Within a selection process, it favored locally a linearly produced output, in which it identified the self and significantly enlarged the consciousness.

In the linearization process, the human brain had reduced the work of entropy, diminishing some of the nonlinear mental activity.

Precuneus is the core element that allowed us to generate a linearly designed civilization.

Mapping the consciousness

Abstract

The aim of this paper, as the title says, is to map the consciousness. It starts with the theories of consciousness, and continues with many subsection, where other correlates are discussed.

It goes to mapping, coupling the state and contents of consciousness, functional "latency", a brain anatomic skull comparison of the Last Clade species, symbolic thinking, entropic brain, pooling effect, brain complexity, and emotional dominance in the brain functioning.

I did not discuss the quantum mind theories because they are too many.

I concluded that the "latencies" in the brain functional and sensorial processing led to decreased brain entropy. A difference occurs between analytical and holistic thinking, where the analytical shows a more reduced entropy.

Interestingly, the increase in "latencies" makes available more details of reality.

The "latencies" manifestation affects brain performance by diminishing the processing speed, while the retarded entropy the latency causes, decreases the intelligent response to environmental stimuli. It may be the case that "latency may affect the general complexity of the brain.

Another conclusion regards the precuneus, which contains spindle-shaped cells distinct from pyramidal cells but embedded in the same cortical layer V (five) as the L5p cells. The L5p are the most indicative about functional consciousness. Such neurons play a unique role in social awareness and interoception. Interoception collects and redistributes information; they compose and decompose. The significant evolution of human precuneus in the last 70,000 years may explain why we took a radical behavioral turn compared to other animals.

In the meantime, Dr. Emiliano Bruner's studies explain that the precuneus differs from one person to another and between races. Therefore, it implies that its configuration modifies among individuals slightly, while differentially evolving. Christof Koch and colleagues found experimentally that in the posterior part of the brain exists a "hot spot" that processes conscious information. This location relates to the cerebellar area that encountered significant development in the last 500,000 years, helping the species of the Last Clade to become distinct from monkeys.

Discussion

A brief review of the consciousness research

Dirk K. F. Meijer and Simon Ragget in their study *Review/Literature compilation: The Quantum Mind Extended* shows the evolution of consciousness research. This article was published in May 2014 by several sources: Academia.edu, Scientific Reports, and Research Gate.

"The study of consciousness was a taboo in academic circles through much of the 20th Century, at least in part due to the long reign of behaviorism. Even the study of emotion being largely proscribed, with brains conceived as being reasoning machines and nothing else."

"The traditional explanation for consciousness or the soul in more traditional language is known as dualism. This posits a separate spirit stuff and physical stuff, with the spirit stuff capable of acting on the physical stuff, as when the soul commands the body."

"Functionalism was at least in the 1990s the dominant explanation for consciousness, driven by the success of computers as problem solving and memory storage machines. The main proposal is that any system or machine that processes information in the same way as the brain will be conscious, regardless of what it is made of. The biological matter and structure of the human brain was deemed irrelevant."

"Identity theory may have been the next most popular theory after functionalism in the 1990s. This declared that consciousness was identical to the brain or identical to its processing."

"Epiphenomenalism was and remains another popular idea. The theory proposes that consciousness is a by-product of neural processing that has, however, no function. The theory also conflicts head on with physics in which there is no acausality with every object or process having influence elsewhere."

"Consciousness resulting from embodiment has been possibly the most fashionable of these ideas. Initially embodiment ideas did represent a genuine step forward in both consciousness studies and psychology as a move away from the brain as a computer in a vat. It is now accepted that mental events could influence the body and visceral events could feed back on the brain. It also accepts that emotion is a relevant aspect of mental life."

"The attempt to classify consciousness as a form of information or information processing has also become fashionable in this century. At a more philosophical level, there is a core difference between information and reality, in that information embraces only what we happen to know, while it can also be defined as an attempt to describe nature's behavior and microscopic make up that comprises reality."

"*A popular but poorly based concept is to call consciousness an emergent property. The idea of consciousness as an emergent property of classically described matter is superficially plausible, and as such can sometimes look like the best shot of modern consciousness studies. The problem for consciousness as an emergent property is no arrangement of such particles and forces has been identified that could produce consciousness.*"

"*In the last two decades, consciousness studies have gone off in a different direction from physics or neuroscience.* ***Many consciousness studies are dominated by philosophers and psychologists who have only a scant interest in what has been happening in brain science, let alone physics. The philosophers are contradicted by the neuroscientist's discoveries***."

Further on, the authors descend into Quantum world and its relationship with the consciousness. "*There is claimed that the brain's neuronal system forms an intricate network and that the consciousness this produces should obey the rules of quantum mechanics*."

"*Quantum mechanics laws are usually only found to apply to very low temperatures. At higher temperatures, classical mechanics take over*."

In Cristiane de Morais Smith opinion (published in INVERSE, on July 25, 2021) "*the **fractals** have been used to explain the complexity of human consciousness. Because **they are infinitely intricate, allowing complexity to emerge from simple repeated patterns**. But if it is the case, it could only be happening on the quantum level, with tiny particles moving in fractal patterns within the brain's neurons. In our latest research, my colleagues at Shanghai Jiaotong University and I went one step further**, being able to reveal the quantum motion that takes place within fractals in unprecedented detail***."

My own approach in this matter tends toward the Science of Complexity and Neuroscience.

* * *

Can states of consciousness be mapped in the brain?

I would like to present in brief a popular science site online, where a group of science contributors participates, and described how *Can states of consciousness be mapped in the brain?*

"*Some neurobiological models of consciousness, such as the Global Workplace Theory, assume that the contents of consciousness are widely distributed in the brain. This assumption has been confirmed by many brain-imaging experiments, those of Stanislas Dehaene and his collaborators. It recorded a major increase in activity in the frontal, prefrontal, anterior cingulate, and parietal cortexes*."

"*Thus, conscious sensory inputs would appear to produce far more extensive brain activity that comparable unconscious stimuli, and a sudden activation of the frontal and parietal lobes would appear to be the typical signature of a conscious perception. But this perceptual consciousness is a "primary consciousness and is not the only form of consciousness."*

"Starting from this premise, authors such as Damasio distinguish a very primitive form of consciousness that he calls the proto-self and that is more like a moment-to-moment perception of the body's internal emotional state. This state is associated with activity of such brain structures as the reticular formation, the hypothalamus, and the somatosensory cortex."

"*The reticular formation is also associated with consciousness in the minimal sense of wakefulness*."

"*It should be noted here that the activity of the reticular formation seems to be necessary but not sufficient for a more elaborated level of consciousness. This later level is attained with what several authors call primary consciousness, meaning a waking state in which we are in relationship with our environment here and now*."

"*Damasio calls this type of consciousness "core consciousness" and says that it depends chiefly on the cingulate cortex and on the interlaminar nuclei of the thalamus.*"

"*Rodolfo Llineas hypothesized that the oscillations of certain neurons in the thalamus serve as a sort of basic rhythm with which the cortical oscillations of the various sensory modalities synchronize themselves to form a unified image of the environment. The thalamus is the place where the signals from all the senses (except small) must pass through it before they can reach the cortex.*"

"These thalamocortical loops have come to play an important role in practically all the neurobiological theories that attempt to explain the higher states of consciousness, for which the lower level of consciousness is in a sense only the prerequisites. These higher levels of consciousness are known as reflexive consciousness and self-consciousness."

"*Three structures-the angular gyrus, the precuneus, and the anterior cingulate cortex may be part of a functional network that makes self-consciousness possible. Among all the areas of the brain that are active during this state, the precuneus is the one that shows the highest rate of neural activity. But it is less active during tasks that make no reference to the self.*"

"*Precuneus seems to play a role in visual/spatial imagery because people have a propensity to represent their own bodies in space."*

Insular cortex seems to map the bodily states associated with our emotional experience, thus giving rise to conscious feelings."

"However, the school of thought known as embodied cognition indicates that ***conscious rational thought cannot be separated from emotions***."

"*Anterior cingulate cortex is an important interface between emotions and cognition, and more specifically in the conversion of feelings into intentions and actions.*"

"*Some neuroscientists, such as V. S. Ramachandran, even suggest that the mirror-neurons ability to decode other individuals' state of the mind may have evolved first, and subsequently been applied to the self., to become what is called self-consciousness. Not only the mirror-neurons, but all parts of the brain that contribute to language, such as Wernicke's area in the temporal lobe, must inevitably play a role in this process."*

"Edelman argue that consciousness cannot be attributed to any specific structure in the brain."

* * *

Coupling the State and Contents of Consciousness

Jaan Aru, Mototaka Suzuki, Renate Rutiku, Matthew E. Larkum, and Talia Bachmann published a study (*Coupling the State and Contents of Consciousness*) in Frontiers in System Neuroscience on 2019 Aug 30 (doi: 10.3389/fnsys.2019.00043).

The authors explain:

"*One fundamental feature of consciousness is that the contents of consciousness depend on the state of consciousness. Here, we propose an answer to why this is so: both the state and contents of consciousness depend on the activity of cortical layer 5 pyramidal (L5p) neurons. These neurons affect both cortical and thalamic processing, hence coupling the cortico-cortical and thalamo-cortical loops with each other. Functionally this coupling corresponds to the coupling between the state and contents of consciousness. Together they form the broadcasting system, where the L5p cells are the central elements.*

This perspective makes one quite specific prediction: cortical processing that does not include L5p neurons will be unconscious. More generally, the present perspective suggests that L5p neurons have central role in the mechanism underlying consciousness."

"However, one of the basic fact about consciousness is that the state of consciousness can never be dissociated from the contents of consciousness. One cannot be conscious of the coconut taste while being in an unconscious state.

There is more evidence showing that L5p neurons are tightly linked to the contents of perception. The researchers observed that the Ca2+ signal of apical dendrites of L5p cells was well correlated with the behavior of the animal. Hence, these findings support the conclusion that L5p cells, especially those projecting to the thalamus, are vital for conscious perception."

The above quoted study makes me think that the L5p cells may have the role of a filter that transforms the nonlinear/multidimensional perception signals into linear outcomes.

I like to quote another study (*Spindle-Shaped Neurons in the Human Posteromedial (Precuneus) Cortex*) of Francisco Javier Fuentealba-Villarroel, Josue Renner, Arlete Hilbig, Oliver J. Bruton, and Alberto A. Rasia-Filho, published in Frontiers in Synaptic Neuroscience on 2022 Jan 11 (doi: 10.3389/fnsyn.2021.769228).

The authors explain:

"*The human posteromedial cortex (PMC), which includes the precuneus (PC), represents a multimodal brain area implicated in emotion, conscious awareness, spatial cognition, and social behavior. Here, we describe the presence of Nissl-stained elongated spindle-shaped neurons (suggestive of von Economo neurons, VENs) in the cortical layer V of the anterior and central PC of adult humans. We discuss the morphology of these cells and those considered VENs in cortical areas forming integrated brain networks for higher-order activities*."

"*In humans, VEMs have been reported mainly in the anterior cingulate cortex (ACC) and frontoinsular cortex, but also in the dorsolateral prefrontal cortex and in the medial frontopolar cortex.; VENs may correspond to cca. 3% of all neurons in layer V in the ACC, being more numerous in the ACC and FI of humans than in apes.*

Human VENs show the feature of excitatory projecting neurons. They may provide fast interconnections between neocortical areas, such as the ACC and FI cortices, and recent transcriptomic data suggest that human VENs may project to extratelencephalic, subcortical targets. They may link emotion and control of sympathetic/parasympathetic sites in brainstream and spinal cord regions."

"VENs are morphologically different from neighboring pyramidal neurons and are larger than layer VI "fusiform" neurons." "The shape of VENs is also characterized by the presence of two main perpendicularly oriented thick primary dendritic shafts, one ascending toward more superficial cortical layers and another descending toward the inner cortical layer."

I quoted this study because the **precuneus is the most recent anatomical development in human brain**, while such development may have differentiated us from any other animal. Thus, its neurons may hide the mystery that led to our modern brain functioning.

Precuneus coupling to other brain areas makes it act as one of the main hubs of the human brain. However, **the spindle-shaped neurons belong to the same cortical layer 5 as the L5p neurons discussed above**. Are they a distinct but novel type of filter of the brain?

Nevertheless, **according to Bruner et al., the precuneus varies between individual humans and among races** that may be indicative of an uncompleted evolution of this section of the brain. It can still evolve slightly distinct from one person to another.

Other experimental studies show that **precuneus influences** an extensive network of cortical and subcortical structures **by producing "associative information" that means it helps compose information by employing a visuo-spatial combinatory process of mental images as they are retrieved from memory.**

In this sense, precuneus is seen as the hidden leader of the novel feature of the Frontoparietal Synthesis created in the last 70,000 years.

Emiliano Bruner, Gizeh Rangel de Lazaro, Jose Manuel de la Cuetara, Manuel Martin-Loeches, Roberto Colon, and Heidi I. L. Jacobs in their study *Midsagittal Brain Variation and MRI Shape Analysis of the Precuneus in Adult Individuals*, published in the Journal of Anatomy (J.Anat., 2014, pp367-376), and online Verlag, 04 April 2013, explain:

"*The major brain covariance pattern within adult humans is strictly associated with the relative proportions of the precuneus. Its morphology displays a marked individual variation, both in terms of geometry (mostly in the longitudinal dimension) and anatomy (patterns of convolution).*

Precuneus does not represent an individual module, being influenced by different neighboring structures."

Maxime Freton, Cedric Lemogne, Loretxu Lehericy, and Philipe Fossati, in their study *"The Eye of the Self: Precuneus Volume and Visual Perspective During Autobiographic Memory Retrieval*" showed that they documented "*individual differences in the precuneus gray matter volume*."

Rosa Li, Amanda V. Utevsky, Scott A. Huettel, Barbara R. Braams, Sabine Peters, Eveline A. Crone, and Anna C. K. van Dujivenvoorde, in their study *Development Maturation of the Precuneus as a Functional Core of the Default Mode Network*, published in the Journal of Cognitive Neuroscience (2019)31 (10): 1506-1519, indicate: "*Our results highlight the distinct role of the precuneus in tracking processing state, in a manner that is both present throughout and strengthened across development*."

Hebscher Melissa, Levine Brian, and Gilbo A. Asaf in their study "*The precuneus and hippocampus contribute to individual differences in the unfolding of spatial representations during episodic autobiographical memory,*" (doi.org./10.1016/j.neuropsychologia 2017.03.029) found that the "***egocentric remembering was positively associated with precuneus volume***."

In other words, precuneus develops and fluctuates along a lifetime due to variations in the gray matter volume, and this contrasts to some other cerebral structures.

* * *

The "latency" involved in Analytic and Holistic thinking.

Yuchun Tang, Lu Zhao, Yunxia Shi, Rui Fang, Xingtao Lin, and Arthus Toga published the study "*Brain Structure differences between Chinese and Caucasian cohorts: A Comprehensive Morphometry Study*" in Human Brain Mapping, 2018 May (doi:10.1002/hbm.23994). it was also published online on 2018 Feb 5.

The authors found that "*The SBM (surface-based morphology) analysis revealed that compared with Caucasians, the Chinese population showed larger cortical structures in the temporal and cingulate regions and smaller in the frontal and parietal cortices*."

"*East Asians, compared with Westerners, process visual information in a more holistic way considering the relationship between objects and context, whereas Westerners focus on salient objects independent of the context in an analytical style*."

"*East Asians showed increased neural activity in the brain regions related to mentalizing and emotion regulation, whereas Westerners culture was associated with increased mental activity in the brain areas related to self-reflection and emotional responses*."

This study indirectly indicates that **a brain with a more developed Frontoparietal cortex tends to increase the analytical activity known as the Frontoparietal Synthesis**.

A Korean study of Dong Woo Kang, Sheng-Min Wang, Hae-Ran Na, Sonya Youngju Park, Nak Young Kim, Chang Uk Lee, Donghyeon Kim, Seong_Jin Son, and Hyun Kook Lim, titled "Differences in cortical structure between cognitive normal East Asian and Caucasian older adults: a surface-based morphometry study", published in Scientific Reports (2020) 10:20905 (doi.org./10.1038/s41598-020-77848-8) has concluded:

"*Significant vertex-wise differences between East Asians and Caucasian groups in cortical thickness occurred. The East Asian group showed higher thickness in the bilateral superior frontal gyrus, caudal middle frontal gyrus, inferior parietal cortex, insula, , inferior temporal gyrus, superior temporal gyrus, middle temporal gyrus, left paracentral lobule, precentral gyrus, fusiform gyrus, right postcentral gyrus, par orbitalis, medial orbital frontal cortex, and lingual gyrus*."

"*We found significantly larger cortical volume in the East Asian older adults compared to Caucasians*."

"*Regarding the superior frontal and middle temporal gyrus, the differences can be explained in part by language differences between ethnicities*."

"***The inferior parietal cortex was found to be thicker and larger cortex in East Asians***. *Also, it occurs an increased cortical density in Chinese speakers compared with English speakers. The prefrontal and parietal cortices have been known to be vulnerable to cortical atrophy. It seems to occur an accelerating atrophy in these cortical regions of the East Asians during the aging process*."

It is known that the brain and body entropy increase with age, causing an increased functional disorder. It implies that the East Asians are more prone to a faster while more local brain entropy increasing with age.

In the meantime, most studies found the East Asian brain volume to be slightly larger than of the Caucasians. It implies more lifetime energetic needs and consumption than for the Caucasians. This fact can be projected in the East Asian aging process.

On the other hand, higher intelligence is also associated with higher brain entropy.

However, the brain development of the last 70,000 years enhanced entirely the precuneus size and function. Therefore, **it may be the case that precuneus is solely responsible for analytical thinking**, which is maybe the only novel function acquired in these 70,000 years.

In the meantime, the development of the analytical function occurred to the detriment of the holistic or more extensive view of reality embedded into the context. This evolution seems to continue nowadays.

For the above-stated reasons, I see the Westerners attempting to globalize their way of thinking as being detrimental for those humans who try understanding the growing complexities we are creating around us.

* * *

The Holistic vs. Analytical Brain Activity

"*The Theory of complexity reveals patterns of evolution in complex self-organizing systems, thus attempting to construct new evolutionary holism.* ***Holism is the very way of analyzing complex systems and determining possible ways of their evolution***."(laco.vencal blog on Feb 14, 2022, published in Smart Cities).

Anastasiia V. Bakhchina, Vladimir V. Apanovich, Karina R. Arutyunova, and Yuri I. Alexandrov, in their study *Analytic and Holistic Thinkers: Differences in the Dynamics of Heart Rate Complexity When Solving a Cognitive task in Field-Dependent and Field-Independent Conditions*, published in Frontiers in Psychology, 26 November 2021 (doi:10.3389/fpsyg.2021.762225) indicate:

"*In general, analytic thinkers tend to focus their attention on an object rather than the field it belongs to.*"

"*Differences in cognitive performance between holistic and analytic individuals have been demonstrated across various tasks involving attention, categorization, causal attribution, tolerance of contradiction. These differences are reflected in problem solving strategies and brain activity of analytic and holistic thinkers*."

"*Holistic thinkers showed significant intersubject correlation (ISC) and more extensive cortical areas than analytical thinkers* (*Mareike Bacha-Trums et.al., 18 Nov 2018*). Significantly higher ISC was observed in holistic thinkers in occipital, prefrontal and temporal cortices."

"*Traditionally, the left hemisphere has been identified to be more involved in processing analytical and verbal information, and the right hemisphere processing holistic, context-sensitive, and spatial information. (Jackson, 1978; Ornstein, 1972; Moskovich, 1979, Springler and Deutsch, 1989; Reuter-Lorenz and Miller, 1998).*"

"*In our study, however, we report contrasting findings, which may have been due to difference in measures (Mareike Bacha-Trums et. Al., 2018).*"

"*In conclusion, this, together with more similar eye gaze patterns in holistic participants, suggests* ***that analytical participants focus on smaller details*** *and are thus more idiosyncratic in their perceptual strategies of perceiving.*"

Manuel Brauch and Andreas Grossler in an economics study (*Holistic versus analytic thinking orientation and its relationship to the bullwhip effect*) published in Systems Dynamics Review/Volume 38, Issue 2/p.121-134 (doi.org/10.1002/sdr.1702) concluded:

"*Holistic thinkers report a higher use of information and that is negatively correlated with bullwhip effect. Bullwhip effect refers to the amplification of demand (order) variability (variance) along supply chains. The bullwhip effect is generally undesirable as it results in excessive inventory, lost revenues, and poor customer service (Carlsson and Fuller, 2000). Our study found that a* ***strong holistic thinking orientation is related to a lower bullwhip effect***."

It indicates in the first place that analytic thinking tends to find more details of the reality. This is similar in many respects to the processing of functional "latency" that leads to a larger number of details extracted from the perceived reality.

In the second place, the result suggests that analytical thinkers may have an "extended latency" compared to holistic thinkers and this is related to their eye movement.

If the bullwhip effect refers to a "variance in the number of details", this study's result confirms that **the holistic thinkers are more meticulous about considering the effect of all together details. At the same time, they opt, during their decision-making, to be quite selective by filtering out many of the details embedded in such subject matter**.

Also, the study indicates that analytic **thinkers use a lower entropy in their decision-making process compared to holistic thinkers**.

* * *

Consciousness association with a functional latency

It is imperious here to specify that my paper did not use Gerald Edelman's concept of primary (defined as limited to present) and secondary (defined as an access to history and plans) consciousness.

Instead, I use Robin Carhart-Harris concept based on experimental drug interpretation. According to Carhart-Harris and other researchers, this consciousness is post-Paleolithic or Late Paleolithic. At the same time, the mentioned concept enhances the entropic contribution to primary consciousness.

Recently in the literature, the "**hot spot theory**" was formulated by Christof Koch et al., which postulates that **the anatomical neural correlates of consciousness are primarily localized in a posterior cortical hot spot in parieto-occipital area**. It is supported by electrical simulation and neuroimaging studies.

The hot spot is also justified by the increase in primary and secondary **consciousness latency, which refers to a correlation of frontoparietal circuits with distal parieto-occipital ones**. The neuroscience studies found that **intuition is generated by the ventromedial prefrontal cortex (vmPFC)**, which can be projected on the forehead, and it is marked in some cultures by a painted dot. It indicates that some unconscious information is processed in the prefrontal cortex, like the vmPFC.

However, the unconscious information from vmPFC travels one of the longest circuits in the brain to **the occipital area where the cerebellum and the "hot spot" are located**. This circuit produces the largest "latency" in the brain's functioning that reflects the consciousness processing.

A group of researchers from Niigata and Kyoto universities, Japan, published in Scientific Reports a study (*Cerebral cortical processing time is elongated in human brain evolution*) on Dec 11, 2022 (doi:10.1038/s41598-022-05053).

The authors explain:

"*There is a sometimes-overlooked disadvantage to having an increased number of cortical neurons, namely* ***a prolonged neural-processing time****. An increase in total number of cortical neurons implies that the number of neurons that support a specific cortical function, e.g., detecting a sound event or identifying a familiar face, also increases. Then, the* ***total neural-processing time for executing a brain function is expected to become longer****.*"

"*This study focused on audition.*"

Networks behaving with latency are those that show low entropy.

In my opinion, the **human brain evolution toward increased functional "latencies" is indicative for a move toward a low-entropy zone**.

Interestingly, **increased "latency" is responsible for revealing many details of the reality**. Comparing monkey latency of 40-100ms to human one that ranges above 200-300ms, while peaking at 1400-1700ms, one can postulate that the **development of "brain's functional latency" was the mechanism that made humans discover much more details of reality than our cousins and any other animals**.

It is known that **other "latencies" are developed by all sensorial systems, including the oculomotor and auditive systems**, and all these "latencies" are much larger than in monkeys.

To these other "latencies" also have been attributed an increased amount of disclosed details of the perceived reality.

* * *

A brain anatomy comparison between Neanderthals, Early Homo sapiens and Modern Homo sapiens

I would like to revisit the changes in the anatomy of the brains of Homo sapiens, Neanderthals, Early Homo sapiens and modern Homo sapiens analyzed by neuroscience and anthropological literature.

I must cite the Max Plank Human Evolution Institute (2021) recent research and that of Bruner et al., (2021). Bruner has written extensively and repeatedly about brain anatomy.

This article was discussed in detail in one of my previous papers.

The Max Plank study refers to marking stages of 100,000, 70,000, and 23,000 years ago. It shows significant anatomic changes occurring between mentioned species during the marking stages.

In short, Bruner et al., focusses on the **exceptional development of precuneus**, which incipiently occurred since 100,000 years ago, while the most evolution took place between 70,000 and 23,000 years ago. This evolution existed only in the MH (modern Homo sapiens).

Bruner indicates that most, if not all, **differences between human adults, including racial differences, are generated by a variation in the development of each individual precuneus**.

I will quote a Japanese study (*Reconstructing the Neanderthal Brain Using Computational Anatomy*) published by Scientific Reports on 23 March 2018.

The Japanese study says:

"*We found that the NT (Neanderthal) had significantly relative smaller cerebellar hemispheres than Homo sapiens, particularly on the right side.*"

"*The superior medial part of the parietal lobule (the precuneus) enlargement in MH (Modern Homo) is much greater than in NT and may have improved cognitive function in harmony with the cerebellar hemispheres and the frontal region."*

"We also found that the relative volume of the right cerebellar hemisphere was significantly smaller in NT compared with EH (Early Homo sapiens), with no difference between EH and MH and no difference in the left hemisphere among the three groups."

"Our findings of laterality in terms of the relatively small right cerebellar hemisphere of NT indicates a minimal connection to the left prefrontal regions, which has one of the major roles in language processing, potentially causing disparity of language ability between NT and Homo sapiens."

The interhemispheric cerebellar connection has been paramount in the later development of language in Homo sapiens around 30,000 years ago.

To recall, EH segregated from NT about 200,000 to 175,000 years ago by the occurrence of the mentioned brain anatomical differentiation.

The no distinction in the left cerebellar hemisphere in these three groups indicates the **absence of a coherent language dominated until 30,000 years ago in MH**.

As it seems, the right cerebellar hemisphere, indicating cognition, has been developed further in EH compared to NT. The same cited study enhances that the **main distinction between monkeys and Last Clade, which included the NT, refers to the significant bi-hemispheric cerebellar development in NT**.

However, the right cerebellar hemisphere development in NT and EH has determined an essential change in their "primary" consciousness development. Such growth continued in MH (starting 130,000 years ago) with additional "primary" consciousness changes, which gradually occurred 100,000, 70,000, and 23,000 years ago. The changes steadily developed in "abstract" thinking, manifesting in artifacts, and produced continual improvements in cognition and memory.

"Larger cerebellar hemispheres were related to higher cognitive and social functions, including executive functions, language processing, and episodic and working memory capacity. Based on archaeological records, Wynn and Coolidge suggested that NT had a smaller capacity of working memory. Thus, the differences in the neuroanatomical organization of the cerebellum may have resulted in critical difference in cognitive and social ability between the two species."

* * *

Symbolic thinking associated with a primitive language is driven by the Cross-Modality Information Transfer

To enhance the development of the symbolic thinking without a coherent language, I will quote a study by Shigeru Miyagawa, Cora Lesure, and Victor A. Nobrega (*Cross-Modality Information Transfer: A Hypothesis about the Relationship among the Prehistoric Cave Painting, Symbolic Thinking, and the Emergence of Language*), published in Frontiers in Psychology on 20 February 2018.

The authors explain:

"We take up the topic of cave art and archeoacoustics, in particularly the discovery that cave art is often closely connected to the acoustic properties of the cave chambers in which it is found. We argue that cave art is a form of cross-modality information transfer, in which acoustic signals are transformed into symbolic visual representations."

"The rapid innovation in tool making and the production of the Blombos Cave ochres and pierced shells suggest that modern humans by around 100,000 years ago could tap some cognitive resources that had not existed before."

"The subject matter and location of the pictures relate directly to the acoustics of the cave structure. Waller (2002) points out that the pictures often cluster in areas with enhanced acoustic properties."

"The auditory representation is mentally transformed into an external, visual representation. This is a pure form of externalized symbolic

thinking, where ***information from one modality transforms into representation in another****. Or externalization may have begun closer to forming the new cognitive capacity*."

The Blombos Cave art of South Africa was dated around 73,000 years ago. The Borneo cave painting dates from 64,000 to 43,000 years ago. Also in Indonesia, the Sulawesi cave (Leang Tedongnge) painting dates 45,000 to 39,000 years ago. The cave paintings in the caves of both sides of the Pyrenees Mountains are dated 33,000 to 30,000 years ago.

The above dates are indicative for a period in which the Homo sapiens started to develop "symbolic thinking", while such new mental ability became stimulated in caves with special acoustic properties, like by the cave hallucinogenic-auditory effect.

The authors of the above study concluded that the unconscious Paleolithic thinking was expressing things without having a language to name them. Similarly, **all other mental symbols represent a domain for which a primitive or raw language cannot namely represent**.

However, besides the very limited language ability for an epoch before 30,000-23,000 years ago, the information presented above gives us a good idea about how the "primary" and "secondary" consciousness of this epoch (70,000 to 23,000 years ago) have been manifested.

There is known a documented opinion that supports that the Paleolithic people have some use of hallucinogenic substances as they accidentally found them in plants during their search for food. Eventually, that era "Venuses" had a similar origin.

Sidath Rankaduwa and Adrian M. Owen in a study (*Psychedelics, entropic brain theory, and the taxonomy of conscious states: a summary of debates and perspectives*) published Neuroscience of Consciousness, Volume 2023, Issue 1, 2023, published on 04 April 2023, explained:

"*One theory that describes how psychedelics influence brain activity is the "entropic brain theory" (EBT), which attempts to understand conscious states-normal and psychedelic-in terms of "brain entropy."*

It proposes that the psychedelic state is itself a "higher" level of consciousness."

* * *

Entropic Brain Theory

"One of the dominant theories in this field is the "entropic brain theory" (EBT), a formal account from Carhart-Harris et al. (2014), which describes consciousness in terms of "brain entropy.""

"EBT attempts to understand psychedelic states in terms of their underlying neural changes-more specifically, how they rank in terms of brain entropy. By "entropy", the authors essentially mean "disorder" or "randomness." To say that conscious states are a function of brain entropy is to say that the level of entropy (or disordered activity) in the brain serve as an index for one's "level" of conscious experience."

Carhart-Harris and Friston (2010) evidenced by insights in cognitive neuroscience, which together suggests that large-scale brain networks (such as DMN) work to predict and constrain the activity of lower-level limbic systems."

"In the context of EBT, primary and secondary states are distinguished in terms of their underlying entropy."

I would conclude with the above quoted study.

The brain's default mode network (DMN) represents a group of functionally related brain regions. It maintains the brain entropy at a level that allows to control the outcomes of behaviors.

One's subjective experience is mainly determined by the level of entropy in the brain. Here, it is to understand that **higher levels of entropy cause greater levels of disorder and uncertainty, but also it increases the level of possibilities, which can lead one to gain insight on new ideas and perspectives**. The primary consciousness shows a great deal of entropy because the entropy suppression was low.

Osvaldo A. Rosso in his study *Entropy changes in brain function (published in International Journal of Psychophysiology, Volume 64, Issue 1, April 2007, page 75-80)* indicates:

"*In present work quantifiers-based information theory and wavelet transform are reviewed. The relative wavelet energy provides information about the relative energy associated with different frequency bands present in the EEG and their corresponding degree of importance. The normalized total wavelet entropy carries information about the degree of order-disorder associated with a multi-frequency signal response.*"

"*The ensuing Shannon entropy-form, based on the wavelet transform, is called the "normalized total wavelet entropy". The wavelet entropy appears thus as a natural measure of order in EEG signals and, more specifically, of the degree of synchrony of the group of cells involved in different neural responses.*"

However, within the primary consciousness, beliefs and habits are more flexible, but at the same time they are more malleable, allowing an easy bending.

As several biophysics study postulate, consciousness appears to **arise naturally because of a brain maximizing its information content**. Normal waking states are associated with maximum values of brain's entropy. Also, the research indicates that the **connectivity itself is generated by the number of different ways in which a certain degree of connectivity would be achieved**.

Many biophysicists believe that **consciousness could simply be an "emergent result" of a system that tries to maximize information exchange that is "entropy."**

* * *

Pooling effect

When entropy (energy) drives the neural signals transmission (information) across lengthy circuit connection, it may lead to **a pooling effect, where one resource would satisfy two otherwise different demands**.

It means that **"pooling effect" smoothers the signal in a way that it can satisfy simultaneously a nonlinear/multidimensional functioning and a planar/linear functioning** because the brain uses both modes. This is somehow **like the Cross-Modality Information Transfer** mentioned in a previous section.

My earlier paper, *Chaos, Nonlinear, and Linear in Human Brain*, discusses which brain structures employ nonlinear, and which ones tend to process information in a linear fashion.

* * *

Brain energy consumption

The Journal of Cerebral Blood Flow & Metabolism, in an article published online 2017 Jun 7 (doi: 10.1177/0271678X17708691) states:

"*In human gray matter, there are 32.8 billion total cells with glial cell to neuron ratio of about 3:2, whereas of the 44.4 billion total cells in human white matter an overwhelming majority are glial cells. Despite these immense differences in the number of glial cells between gray and white matter in the human brain, there is a two to three times difference in CMR. The nonsignaling energy demands in gray and white matter are roughly of the same magnitude. This indicates that the main difference between gray and white matter arises from the higher energy costs associated with neuronal signaling in the cerebral cortex.*

It makes energy cost in white matter increase slowly with rising cortical activity (switching from anesthesia to awake state). Most of the energy in the awake state, as much as 70-75% of the total ATP, supports cortical signaling functions, whereas the remaining 25-30% fraction is attributed to nonsignaling functions. In fact, we estimated that the signaling costs may increase up to 85% in awake and physiologically stimulated conditions.. The housekeeping needs of the network are estimated at 11% for gray matter and 24% for white matter. The remaining resting potential is 20% in gray matter and 58% in white matter."

* * *

Brain entropy, complexity, and intelligence

"*Brain entropy measures the variety of configuration possible within a system, or the number of neural states a given brain can access.*"

Glen N. Saxe, Daniel Calderone, Leah J. Morales published a study in PloS One, 2018 Feb 12, on brain entropy and its relationship to intelligence (*Brain entropy and human intelligence: A resting-state fMRI study*).

The authors have concluded: "*We hypothesize that entropy will be positively associated with intelligence. The relationship was most strongly observed in the prefrontal cortex, inferior temporal lobes, and cerebellum. This relationship between high brain entropy and high intelligence indicates an essential role for entropy in brain functioning. It demonstrates that access to variable neural states predicts complex behavioral performance, and specifically shows that entropy derived from neuroimaging signals at rest carries information about intellectual capacity.*"

Soheil Keshmiri published a study (*Entropy and the Brain: An Overview)* in *Entropy* (MPDI*) on Aug 21, 2020.*

The author concluded: "*The review identifies that complexity, as quantitative entropy, is a fundamental property of conscious experience, which also plays a vital role in the brain's capacity for adaptation. It shows the association of entropy and the brain capacity for information processing.*"

"A number of theses consider complexity as a fundamental property of conscious experience where the evolution of brain's complexity manifests as an entropy-enhancing process."

"*An increase in brain distributed information was particularly observable over medium- and long-range brain connectivity. The most evident distribution of information and entropy occurred during the long-range connectivity*."

"*The presence of an increase in complexity indicates the brain's capacity for transitioning between its different states and networks to promote a greater propensity for information processing. It identified a scale-dependent correlation between complexity and entropy."*

"It attributed various brain-related deficiencies and disorders to the decline of the inherent complexity."

Jonas A. Thiele, Aylin Richter, and Kristen Hilger in their study *Multimodal Brain Signal Complexity Predicts Human Intelligence*, published in eNeuro (2022) (doi.org/10.1523/ENEURO.0345-22.2022) concluded:

"*In sum, our results highlight the temporal and spatial dependency of associations between intelligence and intrinsic brain dynamics, proposing multimodal approaches as promising means for future neuroscience research on complex human traits*."

"*Higher complexity was related to increased information processing capacity (Heisz and McIntosh, 2013). Higher intelligence has been associated with higher complexity*."

"*In sum, our study reveals that individual differences in a person's cognitive ability level are reflected in the complexity of temporally highly resolved intrinsic brain dynamics*."

Most studies found that Shannon entropy provides a way of tracking changes in information content as people shape meaning from data.

Associations between brain signal complexity and intelligence revealed that higher intelligence scores were associated with higher variability in individual-specific microstates.

* * *

A view on psychedelic effects

It is assumed that neurons process information by using a collective interaction. As it appears, the maximum entropy model predicts correlated states in cortical networks, and it can be extended for temporal correlations between states.

Most connections between cortical neurons, as measured by correlations or synaptic strength, seem to be weak, but the multineural firing patterns are abundant. It is postulated that neurons do not need strong individual connections to generate a strong collective interaction.

The maximum entropy model is intended to predict multineural firing at one time only, but the laboratories found that the correlated states can arise across several time bins in vitro and vivo.

Aurora Savino and Charles D. Nichols, in a study on rats and humans published on November 2, 2021, in Journal of Neurochemistry, Volume 162, Issue 1, p. 9-23, showed that "***psychedelic-induced rewiring of gene co-expression networks****, which becomes less centralized but more complex, with an overall increase in signaling entropy,* ***is typical of high plastic systems****. Intriguingly, signaling entropy mirrors, at molecular level, the* ***increased brain entropy*** *reported through neuroimaging studies in humans, suggesting the underlying mechanism of* ***higher-order phenomena***."

"***During psychedelics' acute effects****, leading to the observation of* ***increased variability*** *of spontaneous brain activity* ***and changes in brain networks' connectivity*** *(Preller et al., 2018, 2019, 2020). Using these measurements, the entropy (variability of information content) of brain activity has been quantified and shown to increase upon psychedelics (Carhart-Harris et al., 2014).Interestingly, the acute increase of brain entropy under LSD correlates with changes in personality traits after two* *weeks (Lebedev et al., 2016).*" "*More generally,* ***entropy is a measure of the number of possible states of a system and can manifest at different levels in signals of various kinds, not only from the brain's electrical activity but also from gene expression****. In cell biology, it intuitively quantifies* ***cell plasticity and robustness to perturbations****. Indeed,* ***stem cells have been shown to have high transcriptional entropy that decreases across differentiation*** *(Gulati et al., 2020; Teschendorff & Enver, 2017). And tumors, those resistant to drug treatment, display higher entropy than normal tissue (Conforte et al., 2019; Nijman, 2020; Savino et al., 2020). Both systems are highly plastic, able to adapt to the environment and* ***diversify in response to stimuli****. However,* ***diversity can also increase with the loss of stable organizing principles****, correlating with dysfunctions such as aging (Hernando-Herraez et al., 2019).*"

As this and other research indicate, **when entropy increases, it leads to increased variability**. Higher variability defines high dimensionality and nonlinearity.

Also, this study shows that **high entropy generates high plasticity**, or it implies an increase adaptation to changes in the environment because it offers more choices to consider. It shows that as diversity increases, stability decreases.

As I explained into another paper (*Our civilization originates into a hierarchical regression of vertical complexity*), uploaded on academia.edu, **humans evolved from a high and multidimensional complexity by regressing to a lower complexity that is horizontal, linear, and has low variability. In short, this evolutionary path increased the negentropic behavior, in general, lowering the neural entropy**.

However, as it is postulated in various neuroscience research, the primary consciousness was characterized by increased states of variability, which prevented a mental focusing on more stable solutions.

* * *

Emotional dominance

Recent research unanimously states that emotions still dominate modern human behavior by more than 95 percent.

In my paper mentioned above, I postulated that emotions originate from a long adaptive evolution; therefore, they have been in a high while safe dimensional mental domain, where the systems are stable because the surrounding entropy is high. These high dimensional emotions are transferred, by a dimensional reduction, to the horizontal while linear mental complexity. Hence, within the linear domain, the emotions become subject to Frontoparietal Synthesis that divides and multiplies the resulting divisions. Therefore, the multiplication of the original emotions serves better the multitude of linear mental approaches.

The opportunistic increase of local entropy helps a system to supplement that inner energy consumed by the relationships between its components. The excess energy is expelled by dissipation, while the system enjoys its upgrading that also increases its own complexity. The new energy level acquired by a system requires a new organizational approach among its components that is called self-organization.

Thus, entropy increases are temporal and local occurrences, which result in new stages of self-organization. This process is well evidenced by research on meditation. Meditation drastically reduces the brain consumption of energy that is inflicted in dimensional reduction of high dimensional energy.

Anesthesia studies indicate that during anesthesia the brain glucose metabolism is reduced by 55 percent; during sleep the reduction is 23 percent, and during coma is 60 percent. The reduced consumption represents the amount of energy used for reducing high dimensional perceptual information into three-dimensional, linear mental domain.

Excess energy, which is brought into brain functioning by shutting-off the conscient mechanism, occurs during meditation that shuts-off the self-consciousness activity, too. Hence, during meditation intervenes in energy dissipation that causes new self-organization processes responsible for improving many brain functions, like increased attention.

A similar process, like in meditation, occurs during experimental drug administration. Here, too, there is generated a new self-organization that may alienate the previous brain disfunctions.

I can say that higher entropy inflicted dissipation and caused self-organization processes. Hence, **within the primitive consciousness, the dominant raw emotional content was continually adjusted by the self-organization**; it caused adaptation and survivability to the variability of the environment.

Within the consciousness evolution, there **occurred a switch of adaptation: from self-organization of raw emotions to analysis/synthesis of the emotional content that divided and multiplied the raw emotions into a myriad of secondary emotions**. The increased consciousness enormously augmented the tuning of adaptation capabilities.

As it seems, **such a change in emotional mechanism was imposed by a pure requirement of survival to novel environmental conditions** (manifested as an increasing in cosmogenic ionizing radiation in the form of nuclides, like the C14 isotopes, around 102,000-90,000 years ago, 80-65,000 years ago, and again 42,000 years ago that continued until 34,000 years ago, and which were responsible for many waves of neurogenesis, and a climate cooling 70,000 years ago and again the maximal cooling of the Last Glacial Maximum, starting around 25,000 years ago).

Homo sapiens succeeded to change and adapt their emotional mechanism by developing the novel Frontoparietal area through neurogenesis (as we call it the *rationality mechanism*), while the Neanderthals and other hominins failed that task, turning mentally impacted (mentally ill) by those waves of neurogenesis becoming detrimental, eventually experiencing too many unnecessary brain mutations, and perished.

It was an indirect way of surviving: neurogenesis responded to the ionizing and oxidative effects of C14 isotopes. In all animals, the excess neurogenesis caused olfactory developments; in Homo sapiens the new neurons clustered and produced the Frontoparietal development, while the precuneus became the main actor of this process. It happened that the precuneus turned to play the main filtering and redistributing role of the Frontoparietal Synthesis. It is the place were rationality originates.

However, **rationality is not a component of the new emotional mechanism, but a rule that manipulates the linear emotional content**. Time and again, it filters and redistributes.

Even if I have defined specialized areas, like above, there is known that everything in the brain is process collectively and interconnected.

Most of the mentioned practices and even many others (like coffee, smoking and many plants) only increased, locally, entropy, leading to dissipation, and to novel self-organization processes.

The "primitive" deals with a few but raw emotions, while the "secondary or modern" processes the "primary" emotions in a novel way by dividing and multiplying them.

It is true that the "primary" uses more entropy because this domain is significantly smaller and consumes less energy for dimensional reduction. Instead, the unconsumed energy goes to local cerebral functions, which have an entropy increase, more dissipation and more frequent self-organization processes.

As the research found, most animals consume for their "conscient behavior" around 6 percent of the brain energy. The anesthesia studies on animals and humans disclosed that figure, also indicating that humans use 60 percent of the brain energy for conscient processing.

The modern brain, to develop current consciousness, has reduced the energetic consumption in many cerebral areas by making them more energy efficient.

But not everywhere in the brain this change was possible, leaving some cerebral areas on the edge of stability. And crossing such an edge toward instability, it turned easy and somehow common.

The First (70,000 years ago) and Second (30,000 years ago) Cognitive Revolutions Emergent qualities and quantities as information

Abstract
The aim of this paper is to document the meaning of emergent behaviors and their relation to cognitive revolutions.

In my opinion we have two very distinct cognitive revolutions, one occurring 70,000 years ago and the other one that is more dramatic, occurring 30-23,000 years ago.

The paper analyzes the imaginary constructs, which characterize our civilization, their neuronal origine in the Homo sapiens brain, and their manifestation.

Discussion

I hypothesized that "linearity" developed in human mind at a certain point when several neural anatomical changes occurred in the Homo sapiens brain. Anatomically, such changes developed gradually but in a short time interval 70,000 to 23,000 years ago.

Assumably the rudimentary **"linearity" (before 90,000 years ago) was generated by area MT+, dorsal neural areas V3A, V7, and V4D** (Gils Joost Brouwer, and Raymond van Ee-Journal of Neuroscience January 31, 2007), which assured a limited number of predictable choices. During this era the neural bursts of neurogenesis invaded the hominins brains with large amounts of new neurons, and their **connection developed from one synapse to another, on short distances, have served nonlinear functions and were responsible for "nonlinear-thinking"**; eventually such connections became a chain that served local areas.

However, most "nonlinearity" is produced by the frontal lobe.

A short introduction to human brain neural tracts

Cornelius Weiller, Marco Reisert, Ivo Petro, Jurgen Henning, Nikos Makris, Michael Petrides, Michael Rijntjes, and Karl Egger published in Elsevier on online 21 March 2021 the study *The Ventral Pathway of the human brain: A Continuous Association Tract System.*

The authors concluded:

"*Our data showed that the ventral system tracts, traversing through the extreme capsule, form a continuous band of fibers that fan out anteriorly to the prefrontal cortex, and posteriorly in temporal, occipital cortical regions. In all cases a connection between the inferior frontal and middle-to-posterior temporal cortical regions was identified. In the human brain, this tract connects the language domains of Broca area and Wernicke's area. The differentiation in the three tracts, IEOF, UF, and ECF seem arbitrary, all three pass through the extreme capsule. Our data show that the ventral pathway represents a continuum. The three tracts merge seamlessly, and streamlines showed considerable overlap in their anterior and posterior course. Terminal map identified the prefrontal cortex in the frontal lobe and association cortex in temporal, occipital and parietal lobes as streamline endings. This anatomical substrate potentially facilitates the prefrontal cortex to integrate information across different domains and modalities.*"

"*In both monkeys and humans, cortical regions comprising the object recognition pathway lie directly adjacent to the primary visual cortex (V1) in the occipital lobe, extending progressively into more anterior and ventral portions of the temporal lobe.*" (quoted from Scholarpedia).

"*Physiological studies of cell properties also support the functional distinction between ventral and dorsal pathways. Neurons within the ventral steam (V1 and V2 areas, area V4, and inferior areas TEO and TE) respond selectively to visual features relevant to object identification. Neurons within the dorsal steam (within V1 and V2, areas V3, V3A, and middle temporal area MT, and some adjacent area) respond selectively to spatial aspects of stimuli, such as direction and speed of stimulus motion.*" (Scholarpedia).

"*At all stages, the neural connections are reciprocal (meaning a bidirectional exchange of information) probably the feedback projections play a top-down role in vision*" (as in making selections from a unified input).

"*Thus, an anatomical substrate exists for interactions between the two processing streams, and these might serve to integrate what an object is with where an object is.*" (Scholarpedia).

The quotation from Scholarpedia referred to the processing in the monkeys brain. However, clinical cases suggest that the human visual cortex, like that of the monkey, contains two anatomically distinct and functionally specialized pathways: the ventral and dorsal streams. On the other hand, it suggests **a crucial role of tissue in ventral parts of the temporal cortex for object perception and recognition**.

However, **long-range neural connections in the human brain underpin higher-order cognitive skills (such as planning and emotional regulation) which make humans distinct from the rest of the primate relatives**.

Yan, Yu et al, in an article (*Mapping long-range connections*) published in eLife (May 20, 2022) comments his team's findings during a large study titled *Mapping brain-wide excitatory projectome of primate prefrontal cortex at submicron resolution and comparison with diffusion tractography*, published online 2022 May 20 (doi 10.7554/elife.72534).

The author suggests that "**this long-range neural pathway exists in monkeys, but it connects via multiple synapses instead of a single junction as was previously thought**."

"*We found that vIPFC fiber projections passed through external capsule, claustrum, and extreme capsule and anchored to the middle superior temporal region. Although the trajectory of vIPFC between frontal and temporal regions matched well with the diffusion tractography of IFOF, axonal projections of vIPFC never reached the occipital lobe.*"

This mentioned study shows something that generates a huge distinction between the monkey brain processing and the human brain. It seems that tubular tracts for the transmission of perceptual signal information appeared relatively recently in the section between Temporal lobe and Occipital area. In monkeys, this particular section is created only by multiple synapses.

As it was explained, there exists experimental evidence that synapse-to-synapse transmission of signals generates nonlinear processing and outcomes. Also, the experimental evidence indicates that the longer the range of transmission in the tubular tracts is, it induces more linearization of processing because of longer section of friction of signals with the tract-wall that causes more top-down processes and slows the signal speed.

Research found that the pre-frontal cortex (PFC) is a top-down dissipative producer, which breaks or disassembles wholes into parts, and actively produces linearization.

Broca area is in the left inferior frontal gyrus and Wernicke's are is in the left posterior superior temporal gyrus.

Cerebellar functions in humans are executed via connectivity with other brain centers, such as thalamus, providing substrate for speech control in association with speech planning, initiation, and coordination of speech**. It is a key structure involved in time perception**.

It forms a cerebello-thalamo-cortical pathway and plays a **significant role in speech acquisition during childhood, and it is involved in maturing the speech production**.

The characteristics provided in the above paragraphs make me postulate that a synaptic-linking-connection between temporal and occipital lobes prevent a more comprehensive signal exchange between these two cortical areas and may cause a huge "latency" for signal transmission.

In my opinion, in the brains of the Last Clade's hominins occurred a significant neuronal enlargement of the cerebellum in the era starting 120,000 years ago that increased the "latency" for its connection to temporal lobe, but on the other hand the exchange of information between these two areas increased by developing more synaptic connection.

However, the multiplication of the synaptic connections increased the nonlinear functional outcome, triggering the nonlinear-thinking.

A study on birds found that, when cerebellar lesions occur, the birds song becomes longer, and this effect possibly results from singing more song elements and more slowly (the study was titled: *The Role of cerebellum in learned vocal communication in adult songbirds,* and it was published in Scientific Reports).

In humans, clinical evidence indicates that in patients with cerebellar lesions, like in birds, have fluent speech, but their words will lack meaning. This cerebellar deterioration is termed Wernicke aphasia and causes auditory incomprehension, while the subjects are unaware of their lack of comprehension.

The mentioned damages/lesions to cerebellar area are equivalent to increasing in "functional latency", resulting from large increases in the number of neurons in the cerebellar area. The lack of meaning seems to be the equivalent to nonlinear outcomes, which were incomprehensible for those rudimentary linear minds.

The effect does not allow those subjects to recognize time-intervals in the range of 10-150 ms.

A group of archaeological evidence, indicating abstract thinking, was dated around 70,000 years ago. Abstract thinking is nonlinear thinking. Hence, Homo sapiens brain started to switch to nonlinear thinking sometimes 70,000 years ago.

Masazumi Fuji, Saloshi Maesawa, Sumio Ishiai, Kenichiro Iwami, Miyako Futamura, and Kiyoshi Saito made the study *Neural Basis of Language. An Overview on An Evolving Model*, published in Neurologia medico-chirurgico on2016, April 16 (doi:10.2176/nmc.ra.2016.0014).

The authors concluded: "*In the present model, **language is considered to be processed through two distinct pathways, the dorsal stream, and the ventral stream**. The core of **dorsal stream** is the superior longitudinal fasciculus/AF, which **is mainly associated with phonological processing**. On the other hand, **semantic processing is done mainly with the ventral stream** consisting of the inferior fronto-occipital fasciculus and the intratemporal networks. The frontal aslant trac has recently been named the deep frontal tract connecting the supplementary motor area and the Broca's area and it plays an important role in driving and initiating speech.*"

Evolution of nonlinear thinking.

However, there is evidence that, **around 90,000 years ago**, **it occurred a development of "nonlinear-thinking.**" It was an evolutionary process largely supported by unusually strong bursts of neurogenesis (intermediated by powerful geomagnetic excursions). I would postulate that this era may correspond with the epoch 90,000 to 45,000 years ago that followed Blake, post-Blake, and Norwegian Greenland Sea geomagnetic events. As it appears, the culmination of neurogenesis bursts was achieved during Laschamp (43-41,000 years ago) and Mono Lake (around 34,000 years ago) events.

Neuroscience studies indicate that oscillatory synchrony supports top-down signaling from V1/V2 extrastriata cortex that is a decomposition of the perceived images. It develops a top-down directed beta-frequency synchrony that predicts behavioral context of task-related behavioral information. Contextual-thinking is a fundamental characteristic of nonlinear-thinking. The frequencies involved in this top-down processing were in the range of low-beta 16 HZ.

As it seems, in parallel with visual processing, it occurred an increased processing of the perceived acoustic information within the visual signaling circuits.

Here, the brain was making an energetic choice by selecting for its increased number of neurons more acoustic signals, which required less cerebral/neuronal energy for their transport and processing.

Because of its gradual evolution (90,000 to 23,000 years ago)**, in its beginning, such transition was dominated by optical processing**, **favoring a gradual sliding toward nonlinear thinking**. However, as mentioned, **in parallel but at a lower scale, it occurred a significant increase in acoustic neural processing** intended to compensate the ongoing augmentation in the number of neurons and their respective energetic consumption.

As a consequence of such parallel evolution, many researchers found **a threshold between static and dynamic phases of human language, and it is placed sometimes between 50,000 and 30,000 years ago**. Even then, the temporal lobe, responsible for language, did not seem to have evolved much until 35,000-23,000 years ago.

Characteristics of the evolving nonlinear-thinking

The hominin societies would view time as something that doesn't necessarily flow in a straight line from the past to present and future. Instead, the events would be interconnected and co-occurring, everything happening at the same time that would be the present.

Obviously, those Late Paleolithic Homo sapiens societies had adjusted their cultural practices and beliefs to the new modality of nonlinear thinking. It became evident to them that most observable phenomena were cyclic and made them adopt a cyclical belief, where a time cyclicity was assumed.

Probably, a more focused view in archaeology may help find evidence about the approximative eras when the mentioned **transitions from simple or primitive linear processing (that was intended for survival and was common to all species) to nonlinear-thinking occurred**.

The archaeologists may find material evidence about **behavioral changes, which altered the understanding of hominins on memory and cause-and-effect changed**, and this novel but confusing change, provided too many probabilities at once. It increased the uncertainty, and rapidly diminishing the surviving capabilities of Homo sapiens and other hominins.

Some of the changes were favorable, like cyclicity of events, contextual and intuitive thinking, Others aspects fall outside the original linear perception that was inherited, and **caused mental conflicts, like the negation of urgency, and a certain confusion about when things must be done**.

No referential points were allowed, and no compartmentalization, division, and multiplication were possible, which all altered the previous linear skills.

In my previous papers I hypothesized that a chain of **geomagnetic events (geomagnetic excursions) favored ROS production in the hominin brains that stimulated bursts of significant neurogenesis and biophoton production, which were associated with extra but a significant neuronal production**.

Too much neural renewal, after a certain positive accumulation, had caused too many unwanted neural mutations, which generated mental diseases.

Linearization of thinking

In the modern brain (after 70,000 years ago), the increased "linearity" was provided by a significant augmentation in parietal areas and precuneus. It gradually maximized the predictions toward a single choice that occupies a position on a straight line, and all maximized choices became sequentially, while "linearly" arranged. **Those points**, as graphic representation of choices, **fundamental property consists in unlimited additivity or sequentially, which correspond to the process of superposition**.

(The superposition principle states that the net response caused by two or more stimuli is the sum of the responses.)

In my hypothesis**, Home sapiens brain**, when crowded by unwanted mutations, **adapted toward a distinct neural solution** (that gradually developed the **Frontoparietal Synthesis**), where the **anatomical change favored the development of "linear-thinking**."

The extinction phenomenon, which affected the rest of hominins, is dated **45,000 to 30,000 years ago**, and this threshold may correspond to the mentioned **transition from nonlinear to linear thinking**.

Linearity is something that is perceived by the human senses as a referential point and this process occurs between those points placed along a line and where everything must be only sequential. It allows one to add, subtract, multiply, divide, and do most of all simple operations we currently do in mathematics.

The most significant achievement of linearization turned out to be the process of quantification intimately connected with having linear referential points.

Linear characteristics

Biologically, the nonlinear and linear processing are generated by two distinct neural signal transmission modes:

-when the transmission is on **short-distance (nonlinear signal processing)**

-when the **long-distance neural tracts are developed by the brain (it causes signal linearization processes)**

As the mathematics explain, a system is linear when it follows several principles:

-it obeys superposition (additivity) principle

-homogeneity property (scaling), where a zero input yields a zero output

-the system is time-invariant, when a shift in the input signal generates the same time shift in the output signal

-stability where the output of that system is bounded any bounded input.

Linear systems are a singularity in the infinite set of possible systems; the rest of the systems are nonlinear.

Thus, analyzing nonlinear systems without the need for linearization still implies restricting it to time-invariance, restraining it to only a few classes of nonlinearity, because otherwise it will occur an infinite number of sinusoidal signals.

However, nonlinear in mathematics was considered simultaneously with linear: quadratic equations (nonlinear) have been solved four thousands of years ago, and differential equations (nonlinear) were in use three thousand years ago.

There is mathematical evidence that nonlinear thinking inhabits human mind for a very long time, and in my opinion it occurred long before the linear-thinking.

Quality and quantity as information

The quantities at the conscious level are subjective amounts of information generated by the complexity of elements. The quality of the experience is defined by an informational relationship.

The qualia is the term employed to define the “quality” of experience. However, each sentiment has a particular quantity of information, and each experience produces what we call a sentiment.

Quantity is a principled way of transforming qualitative properties of experience into mathematics by employing a point of reference and making the comparison to that chosen point. We compare such quantities of information even when identical information is produced by very distinct sources.

The qualities represent the amount of information that is generated by a system in a particular state. This state is produced by the interaction that occurs among the elements of that system. By comparing different such states, as they complexly evolve, one hopes to disclose the sense taken by the interactions that is embedded in the emergent result.

Let's take the example of the measurement of a day. It must reflect the spin on which our planet evolves every day around its rotating axis. This cyclic movement is approximately 24 hours per day.

As it seems, this rotation is constant and then, it is considered a physical quantity. In fact, our planet rotation is slowing slightly with time. "*Atomic clocks show that the modern day is longer by about 1.7 milliseconds than a century ago.*" (Wikipedia). "*Over millions of years, Earth's rotation has been slowed significantly by tidal acceleration. Some models suggest that Earth maintained a constant day length of 21 hours throughout much of the Precambrian.*"

"*Some recent large-scale effects (like large earthquakes) also contributed to shortening the length of the day by 3 microseconds. The glaciations and post-glacial rebound have their influence in our planetary spin. Large water storage, like the Three Gorges dam, increased the length of the day by 0.006 microseconds.*" (Wikipedia)

"*Random fluctuations due to core-mantle coupling having an amplitude of 5ms. The sideral day is shorter than the stellar day by 8.4 ms.*" (Wikipedia).

Hence, these variabilities are too small to make a practical difference. Thus, the length of the day remains a physical quantity of practical use.

However, different parts of the cortex influence different qualitative aspects of the perceived consciousness. This makes the case of *integrated information* that is generated by every complexity that causes interaction between its components. *Information is defined as a reduction of uncertainty.* When one sees an image, that image is experienced as an integrated whole and cannot be divided into the images of its components, which may be experienced independently.

Integrated information

The Theory of Information suggests that the quantity of a physical system depends on its existing state. The same theory claims that the **quantity is determined by the amount of integrated information**, while the **quality is determined by the set of informational relationships its mechanism generates that is the emergent**.

David Balduzzi and Giulio Tononi published a study *Qualia: The Geometry of Integrated Information* on PLOS Computational Biology on 2009, Aug 14 (doi: 10.1371/journal.pcbi.1000462).

The authors explain: "*In short, generating a large amount of integrated information entails having a highly structured set of mechanisms that allow us to make many nested discriminations (choices) as a single entity.* ***Each of the nested choices is an "informational relationship"****. According to IIT, these mechanisms working together generate integrated information that determines the quality of experience.*"

"*Qualia space (Q) is a space where each point is a probability distribution on the possible states of the system. Quantity is given by the amount of integrated information a system generates.*"

"*Prior to considering its mechanism and current state, the system of two binary elements could have been in any of four possible states with equal probability. This potential repertoire (or apriority repertoire, is the maximum entropy (maxent or uniform) distribution, which entails maximum ignorance.* ***The mechanism and current state of the system, however, reduce uncertainty, i.e. generate information, about the previous amount of the system. This is because only some previous states could have led to the current system state***."

The above concept indicates how compatible the current state is with previous states and allows a comparison between states.

The illusion of reality and its physical background

Cengel, Yanus A. (Department of Mechanical Engineering, University of Nevada, Reno, NV, USA) published a study On Emergent Quantities, Mental Perception and Constructs, and Agencies, subtitled A Holistic View of Existence in The Journal of Neurobehavioral Science on June 12, 2021 (doi: 10.4103/jnbs.jnbs_33_21).

The author explains: "*There is a growing tendency to view all nonphysical quantities as properties of organization and label them emergent quantities which appear out of nothing and disappear into nothingness. In this paper we present plausible arguments that emergent quantities are not limited to large assemblies of matter and that they are encountered even at subatomic level. We also make a distinction among different kinds of nonphysical entities such as mental constructs invented by the mind and entities discovered by inference, such as the laws of nature.*"

"*Science is coding and transposing actual structures of physical realm perceivable by the five senses into representative abstract or nonphysical structures of phenological realm perceivable by the mind. The new insights gained by the manipulation of these abstract structures, including mathematical representations, are then decoded back into the physical realm.*"

"*The appearance of emergent qualities is contingent upon the existence of the physical entities on which they emerge since, otherwise, we cannot perceive the emergent qualities. This sets the stage* ***for the perfect illusion that emergent qualities originate from the physical entity***."

Let's see how Yunus A. Cengel defines some nonphysical entities as emergent qualities, mental constructs, and agencies. Most of these abstract structures are mentally manipulated and decoded back into the physical realm.

"*We are granting nonphysical entities, including emergent qualities and mental constructs the status of existence though not necessarily external existence but discoverable by the mind.*"

"*Most physical properties such as shape, color, hardness, conductivity, taste, and sound are also nonphysical. We commonly view them as being physical things and change when the physical thing undergoes a change.*"

"*Illusions, hallucinations, imaginations, and conceptions are all nonphysical entities by definition since they are not made of matter-energy. Nonphysical things are not conserved. Nonphysical existence can only be logically reasoned and inferred.*"

"*The physical and nonphysical realms are intertwined, and they influence and interact with each other. Scientific inquiries may have profound ramifications on the nonphysical realm, each as emergent properties.*"

"*Many properties or features of physical entities are subjective since they are emergent properties and are not reducible to matter-energy. All physical beings come packed with some intrinsic subjective qualities that cannot be traced back to the properties of their parts.*"

"***All physical existence is in fact a combination of physical and nonphysical realms**, giving rise to the phrase "whole is not only more than but very different from the sum of its parts.*"

"*Life does not qualify as physical entity, but living beings do since they have physical bodies made of mass-energy.*"

"*Emergent qualities manifest on physical entities and the mental constructs emanates from a mind as the renderings of the mind. Emergent qualities and agencies are perceived by the mind, but they are not created by the mind. Mental rendering of perceptions should not be confused with mental creations such as imaginations, dreams, plans, thoughts, and designs.*"

"Meaning, which is the essence of information and represented by symbols such as words, shapes, and O's and 1's is also another form of nonphysical existence."

"*Danish author T. Norretranders said: "The visual world, what we see, is an illusion, but then a very sophisticated one. There are no colors, no tones, no constancy in the real world, it is all something we make up.*"

"***It is important to note that emergent properties passively qualify physical beings and not actively govern them. They do not exert an influence on physical beings**, such as forcing them to act in a certain way. Emergent properties are just qualifiers or qualities and not agencies with causal power.*"

"*Emergent qualities are not mental creations since they are intimately tied to physical entities they appear on in a certain way, and they are not inventions of the mind. They existed long before the mind did.*"

"***All properties*** *as shape, dimensions, volumes, temperatures, pressure, density, entropy, enthalpy, amount of matter, amount of energy, taste, small, hardness, thermal conductivity, electrical resistivity, absorptivity, transparency, molecular structure, and wavelength*-***everything other than the elemental particles of physics themselves and their assembly-exist as abstract entities in conscious mind. All properties are abstractions or conceptualizations***."

"***The characteristic properties of the fundamental particles such as charge, and spin appear to be emergent since they do not originate from the quantized excitations of the underlying elusive fields. Therefore, emergence should be recognized as a fundamental phenomenon, and the nonphysical emergent existence should be regarded as an essential constituent of the realm of existence***."

"*We emphasize that emergent quantities are not mental constructs created by the mind. They would still exist out there independent of the mind. The mind only renders a factual description and mental image of emergent qualities during scientific inquiry.*"

"*It appears that* ***mental constructs or mental rendering constitute a major part of the nonphysical realm of existence. All abstract mental constructs such as mathematics, languages, literature, philosophy, thoughts, beliefs, plans, imaginations, and dreams are part of the nonphysical realm***."

"*Some mental constructs as the laws of physics and mathematics are discovered. They already exist out there unnoticed, until a conscious mind become aware of them and express them using physical symbols of representations such as words and numbers*."

Brain entropy

The brain generates consciousness by reducing the entropy of its internal representations below a critical threshold. This is so, because the brain gradually minimizes the entropy of internal representation and as a result, it creates minimum entropy representations termed "qualia." The QBIT theory indicates that qualia are maximally entangled, and maximally coherent systems with a high amount of information.

The axon-dendrite polarity is required for directional information

The role of new neurons

Newly generated neurons by processes of neurogenesis migrate through the subventricular zone (SVZ) and intermediate zone (Z) to reach the cortical plate (CP). The neurons, which are generated later in brain evolution during life, migrate past neurons generated earlier and occupy more superficial positions.

The cortex is a six-layer structure that is unique for mammals. As explained, neurogenesis places its new neurons on top of the old ones, and a correlation between all these layers always exists.

The connection between neurons occurs via synapses.

A hypothesis on possible anatomical changes in the Homo sapiens brain, which may have triggered a major cognitive revolution.

Archaeological evidence suggests that around 150,000-125,000 years ago occurred a significant enlargement in the skull of all hominins. Such enlargement would indicate an increase in the number of neurons.

Alexander Maier of Vanderbilt Brain Institute published in CellPress on August 17, 2020, an article titled Visual perception*: Human Brain Cells Cause of Change of View* (doi:10.1016/j.cub.2020.06.054).

The author explains recent findings of De Jong et al. as *Endogenous (illusory) perceptual reversals evoke a reverse hierarchy pattern of cortical responses*.

The article describes De Jong's experiments and their results. It indicates that, during the experiments, it was found that it occurred an inverse temporal lobe response preceding those of the occipital cortex. Such hierarchical inversion was generated by an increase in functional latency of the occipital area. As a result, the response of the visual signal in the temporal lobe occurred after the occipital lobe response that was in opposition to established hierarchical functioning. Consequently, such inversion created a holistic neural vision.

In my opinion, the mentioned experiments and results are an indication that the hierarchical inversion was an anatomical change that developed as an outcome of a significant increase in the number of neurons in the occipital area in the epoch 125,000-70,000 years ago. The increased number of neurons produced a functional latency in the transmission of perceptual signals compared with temporal lobe response.

In my opinion, **the significant neuronal increase in the occipital area created the posterior cortical hot zone that is associated by Kristof Koch with the minimal neural substrate essential for conscious**.

Hence, the maturation of the posterior hot zone probably occurred around 70,000 years ago and generated a two-fold outcome:

-it introduced a mental holistic processing

-it caused the mentioned development of the minimal neural substrate for conscious.

Wikipedia explains the role played by the posterior hot zone: "*The sensory component (shape, color, texture) of each object stored in memory is physically encoded by neurons of the posterior cortical zone in structures called neuronal ensembles (NEs) or cognits. When one recalls any object, the object-encoding neuronal ensemble activates into synchronous resonant activity, and it is known as the Binding-by-Synchrony hypothesis. It was proposed that synchronization is a general mechanism underlying any novel imaginary experience.* ***When the synchronization of independent object is driven from the front by the lateral prefrontal cortex, we refer to it as Prefrontal Synthesis; when the synchronization is driven from the back, we refer to it as dreaming or hallucinations***."

The above explanation quoted from Wikipedia suggests that De Jong experiments demonstrate **a holistic, while nonlinear reorientation intervening in the Homo sapiens brain, and which started to manifest as abstract-thinking around 70,000 years ago**. In the meantime, **this alteration stimulated an increase in conscious thinking that was an increase in the previously developed simple or primary linearity**.

Thus, the Homo sapiens brain had turned to a dual processing, where the holism (nonlinearity) was the novel one, while the evolving of conscious was additional to the pre-existing occurrence.

Even then, the novel holistic processing influenced the conscious processing, too. "*Holism of systems and processes make them interact with themselves and produce themselves from themselves.*" (A. Gupta, 2008). It is about addressing the properties of the concrete object (physical object) and selecting the relevant ones. It generates a summation of properties or a reduction of them.

Abstract is a summation of details; non-abstract is detailed information. Holistic refers to a process and a representation of the stimulus. The components are integrated to give rise to a global percept. There is neural evidence that faces in high level visual areas are represented holistically in monkeys infero-temporal cortex (Logothetis & Sheinberg, 1996; Tanaka, 1996) and in human fusiform gyrus (Schiltz & Rossion, 2006).

This is what Dr. Carhart-Harris calls the primary conscious, and which probably existed from 70,000 to 30,000-23,000 years ago.

First Cognitive Revolution

In my opinion or hypothesis**, the First Cognitive Revolution (70,000-23,000 years ago) overlaps the epoch when the Primary Conscious was developed**. As mentioned above, it implies a holistic approach or nonlinear thinking. It was triggered by an occurring functional latency in the transmission of optically (visual) perceived signals within the occipital cortex area, namely the occipital hot spot.

Prof. Yuval Noah Harrari of the Hebrew University of Jerusalem wrote in Naturalmoney.org an study *A Brief History of Humankind.* The first part that I will quote here refers to The Cognitive Revolutions. In this and other of his work he theorized the First Cognitive Revolution.

"*At the time of the Cognitive Revolution planet Earth was home to about 200 genera of large terrestrial mammals weighting more than 50 kilograms. At the time of the Agricultural Revolution, only 100 genera survived.* "

"*Under identical ecological conditions might have been very different societies, cultures, and beliefs, thanks to novel ability (language, too) to create imagined realities. There has been not a single natural way of life.*"

"*Ever since the Cognitive Revolution (First Revolution in my opinion) about 70,000 years ago,* ***sapiens have been able to change their behavior quickly without any need of genetic or environmental change. Sapiens can change the social structure, the nature of their relations, economic activities within a decade or two, just by telling new stories or changing the previous stories.***"

Within a new mode of abstract thinking, a primitive while static language still helped in story-telling, making the sapiens overpass the slow genetic evolution.

"*The* <u>*ability to refer (later has been the a dynamic speech) to things that do not exist, enabled sapiens to cooperate flexibly in large numbers*</u>*. Since* ***large scale cooperation between sapiens is based on stories****, the way people cooperate and build their societies can be changed quickly by changing the stories.*" Here the author emphasizes the role of speech in proliferating the story-telling, but this full ability, in my hypothesis, appeared only during the Second Cognitive Revolution that manifested around 30,000-23,000 years ago.

Several linguists from South Africa consider that the static language begins to manifest simultaneously with increased abstract thinking, while both appeared about 70,000 years ago.

Thus, **the First Cognitive Revolution was triggered and largely benefited from the development of static language that significantly enlarged social cooperation, from development of the Primary Conscious, and from Holistic Thinking**.

All these three factors contributed to an increased brain plasticity, allowing sapiens to move beyond their traditional spaces due to increased adaptability to changing environments.

During this First Cognitive Revolution, sapiens were able to migrate out of Africa populating Eurasia, Australia, and a good part of Oceania just with a static language and a primary Conscious. "*Because of the fastness of cultural evolution, Homo sapiens soon outstripped all the other human species and all the other animal species in its ability to cooperate and to accumulate power.*"

"*None of these things exist outside the imaginary stories that people invent and tell one another. There are no nations, corporations, and money. There are no such thing as human rights and there are no laws, and there is no justice anywhere outside the common imagination of sapiens and the stories which we tell each other.*"

"*The truly unique feature of human language is the ability to transmit information about things that do not exist at all. Fictive language is so important because it enables sapiens to imagine things collectively.*"

"*Beginning 70,000 years ago, there is the first evidence for art and jewelry and the first evidence for trade between different bands. There is also the first evidence for complex societies comprising hundreds of people and not just dozens of people.*"

"*Cognitive abilities are the abilities to communicate, remember, learn, and think. Such things were very limited 100,000 years ago in Africa.*"

About 70,000 years ago appeared the first needle, allowing sapiens to make cloths. Some advances in tool making also occurred at the same time. "*One of the most important technologies that started appearing 70,000 years ago is probably boats and other kinds of sailing craft.*"

Currently, the archaeologists seem shocked by some incomplete evidence that sapiens may have sea fared toward the Americas around 50,000-45,000 years ago, contemporarily with their sea faring from Southeast Asia toward Sahul and New Guinea.

As a result of sapiens migration out of Africa, **it occurred some sexual encounters between Homo sapiens and species of humans and they produced fertile offspring. This is a fact genetically proved, and it qualifies for the rising a new species of sapiens, where Neanderthal**

and Denisovan mixing is quite small (2% to 6%) but probably quite influential.

The genetic impact of Denisovans mixing seems to be larger in East Asia, New Guinea, and Australia; the Neanderthals genetic impact in Europe varies from a region to another and from one individual to another. Last Denisovan disappeared in New Guinea only 15,000 years ago, and the last neanderthal in Eurasia disappeared 30,000 years ago.

Because of the above-mentioned genetic impact, the resulting sapiens-hybrid appears to reflect distinctively this genetic admixture. Even then, in my hypothesis, the sapiens hybrids gradually homogenized all over Eurasia and Oceania (Australia included) with each other and with sapiens who have not been affected. Also, a similar or at least partial homogenization was caused by African back-migration of sapiens.

The revolution in sapiens cognitive evolution, during the First Cognitive Revolution, has occurred almost entirely outside Africa.

The genetic mixing produced the hybrids who have a sapiens genetic base and genetic dominance.

However, during the Second Cognitive Revolution, it occurred several other neural anatomic changes, which together produced a novel linear thinking.

Robert Rosenbaum and Cody Baker published a study *Nonlinear stimulus representations in neural circuits with approximate excitatory-inhibitory balance* published online 2020, September 18 (doi: 10.1371/journal.psbi.1008192).

In their study the authors explain how linear neural circuits turn to produce nonlinear functioning and computations.

They concluded that "*balanced network models predict a linear relationship between stimuli and population responses. Every balanced network architecture admits stimuli that break the balance state and push the network into a "semi-balanced state" characterized by excess inhibition in some neurons.. The semi-balanced state produces nonlinear stimulus representations and nonlinear computations*."

"*The balanced state arises naturally in large scale computational models of neural circuits. The balanced state encourages simple, linear relationships between stimuli and neural responses. Classical balanced state is fragile and easy broken in a way that produces a new state called the "semi-balanced state", where the representations are nonlinear, and improve the network's computational power.*"

Dr. Rosenbaum discovery is paramount for my hypothesis of an intervening nonlinear thinking in human evolution, and which started to evolve in the most hominin brains around 70 ka. It suggests, in my opinion, that previous to 70 ka, it existed a primitive linear computation in hominin brains, because it "*arose naturally and encouraged simple linear relationships between stimuli and neural responses*. (Rosenbaum, 2020).

In my opinion, the balanced state was broken in the era 100 ka-70 ka, and a nonlinear neural representation and computations gradually developed in hominin brains until 30 ka-23 ka. In my opinion, the triggering element was the excess production of ROS and neurogenesis, and such a additional production of neurons implied higher energetic demands. The higher energetic demand broke the balanced state that became replaced by a simi-balanced state, where the arising nonlinear representations and computations improved the computational power because they consumed less energy.

In the meantime, it seems, the development of nonlinear representation and computations enhanced the role played by the short-range circuits, which favor nonlinear processing.

Now, let's see the role played by short- and long-range circuits.

Simon Arvin, Andreas Nargaard Glud, and Keisuke Yonehara published the study *Short- and Long-Range Connections Differentially Modulate the Dynamics and State of Small-World Networks*, on 25 January 2022 in Front. Comput. Neurosci. 15:783474 (doi: 10.3389/fncom.2021.783474). The authors concluded that "*the topological and dynamical properties of the small-world network are divided into two functional domains based on the range of connectivity, and that these domains play distinct roles in shaping the behavior of the critical state. We demonstrate that short-range network connections shape the dynamics of the system, e.g., its volatility and metastability, whereas long-range connections drive the system state, e.g., seizure.* ***These findings lend support to combinatorial neuromodulation approaches that synergistically normalize the system dynamic while mobilizing the system state***."

The mentioned study implies that the system dynamics can produce nonlinear computations, while the system state can remain generally linear.

Another study by Gregory Z Tau and Bradley S. Peterson titled *Normal Development of Brain Circuits*, published on 23 July 2009 in Neuropsychopharmacology REVIEWS (2010) 35,147-168 show that "*alternatively, neural circuit may comprise a network of interconnected brain regions that together integrate vast amounts of information and perform more complicated cognitive and regulatory functions. Clearly, these distributed neural circuits are neither present at birth nor are they invariant through life. The structure and function of neural circuits perpetually changes and evolve from the time of the first contact between the nerve cells. The interplay of inherited genetic programs with a wide range of environmental exposures and experiences determines the birth, death, and cellular characteristics of neurons, as well as the formation and reformation of their axons, dendrites, and synapses*."

"*Inherent to both the construction and maturation synapses within neural circuits are their continuous refinement and modification. The synaptic connections among neurons early in development are often transient stepping-stone toward the more stable connections that characterize more mature circuits*."

"*Local cortical connections appear before more distant ones. Thickening of cortical layer II., which makes short-range cortical connections with other cortical regions, precedes thickening of layer III, which makes long-range connections with other regions, between the age 15 and 72 months.*"

"Overall metabolism rises to twice that of adult level by 4-5 years of age and remains high until 9-10 years of age."

This study pinpoints to the fact that the **development of the architectural brain organization dictates the generation of short-range circuits before the long-range ones. They define the system dynamics**. Only after the dynamics are defined, does the brain architecture allow the development of long-range circuits, which define the system state.

However, **in this childhood phase of development (for 4-10 years old) the metabolism is twice as big compared to an adult metabolism**. Because those children body size and weight are relatively small, it means that a huge part of the energy produced by the body metabolism is dedicated to the brain functioning that is three time more energy (60%) compared to an adult consumption of 20% of body metabolism.

One explanation for this high energetic consumption refers to many inefficient circuits, which cause high dissipation rates. As the study says, the neurons and circuits are continually adjusted and optimized over the lifetime.

A study by Gabrielle J. Gutierrez, Fred Rieke, and Eric T. Shea-Brown, titled *Nonlinear Convergence boosts information coding in circuits with parallel outputs*, published online 2021 Feb 15 (doi: 10.1073/pnas.1921882118) shows that "*computation in neural circuits relies on a common set of motifs, **including divergence of common inputs to parallel pathways, convergence of multiple inputs to a single neuron**, and **nonlinearities that select some signals over others**.*"

*"Our results suggest that neural circuits may preserve more information using suboptimal components. We find that **a convergent circuit with divergent parallel pathways can encode more information with nonlinear units** than with linear subunits*."

This study suggests that the neural network uses divergence and convergence of the perceived signals to activate parallel neural circuitry, while the nonlinear processing encodes more important information than linear processing.

Contemporary nonlinear thinking of the First Nations around the globe.

I would like to provide some examples of nonlinear thinking systems, which persist at present among individuals who have been subject of very long-term isolation. We call them the First Nations or aboriginals because they still possess cultures preserved from prehistoric times.

These areas of previous isolation refer to Australia, Papuan-New Guinea, Amazonia, South Africa, North America, but they also exist in spots located in the jungles of Malaysia and Indonesia.

Sara Kianga Judge of Australian Museum is an aboriginal of Australia and in an article she wrote explains: "*First Nations people don't always think about things in timelines with set starting and finishing spots. Instead, we think in patterns and cycles using the place and time where we are now as a starting point. Instead of a straight line, you now have something more like a circle. This is the simplest kind of nonlinear time-the start and end are no longer deciding the way the story is told.*"

"*More complicated nonlinear thinking doesn't have a set starting or ending at all. The story can be told in any direction and from any time. Instead of lines and circles, nonlinear thinking can create beautiful webs of all different kind of connected patterns that reflect the connected patterns in our lives and ecosystems.*"

Donald Fixico (2003) stated that "*generally, in Native American thinking, a circular philosophy notes that all things within the circle are related and should be equally respected. The talking circles became a dialectic process in which all participants in the circle shared their knowledge without judgement from the other participants.*"

"*Indian thinking is inquiry into relationships and community, and it reminds us that community extends beyond human relationships.*" *It occurs the understanding of the world in which entities are defined not by their material form but rather by their relation with other surrounding entities, knowledge creation is based on investigating how different entities interact with one another. One of the aspects of indigenous knowledge shared across multiple cultural contexts and landscapes is the understanding of circles and placing things within circles, as conceptual framework for marking meaning.*"

A group of French scientists led by Dr. Stanislas Dehaene investigated numerosity among the Munducucu's villages of Amazonia. They concluded that those villagers exhibit a great comprehensiveness of nonlinearity. They observed that linear thinking is better developed in elderly than in youngsters. They found that exists a simultaneous presence of linear and compressed (algorithmic) mental representations of numbers, that is not unique to the Munducucu, and which evolve in parietal and prefrontal cortex.

Regarding the modern population with local roots from Asia-Pacific region, the prehistoric Denisovans seem to be a major influencer of their genetics. Denisovans were a hominin branch that extended from Europe to Altay and from there to East Asia, reaching Papua-New Guinea and Australia. Their genetic influence was also found in South America. Recent evaluations indicate that Xianhe (Gansu, China) fossils were Denisovans or a specimen similar to them. The Xianhe's individuals dated 165 ka, had a cranial capacity of 1,800 cc.

In north China, in Harbin region was found the Harbin men (also named Homo longi and the Dragon Man) and was dated 146 ka. His cranial capacity was close to 1,800 cc.

Genetic analysis shows that Denisovans were more related to Homo sapiens that with Neanderthals.

Most specialists assume that neanderthals and Denisovans were inhabiting the territories were Homo sapiens spread later. Interestingly, all specimens have genetic traces, which combine the Neanderthal and Denisovans inheritance.

I would say that the process of encephalization represented an increase of primitive linear thinking and evolved until nonlinearization started to manifest 70,000 years ago by producing abstract outcomes.

However, the nonlinearization that occurred among all hominins produced a diminishing in skull size because the nonlinear brain is more energetically efficient.

The unconscious is 95-97% of the brain activity, it is nonlinear, and consumes only 50% of the energy of a walking brain. By contrast, consciousness consumes 50% of the brain energy but it can process only 3-5% of the perceived information.

I would refer to several studies which indicate that an adjust IQ test for the age of 5-7 years old found that the most of the tested children have an IQ of 130-150 points. When the test was repeated on the same children after their grade school graduation, their IQ dropped to regular levels of 96-100 points.

Further linearization of children thinking is caused by schooling and stimulates, too, the development and use of long-range circuits. The conclusion here suggests that nonlinear thinking produces much higher IQs.

As one can observe, the benefits of previous nonlinear thinking did not produce a somehow expected social and/or technological progress. The explanation may be found in the association of nonlinearity with previous primitive linearity. Hence, that past combination primitive-linear-nonlinear was not as successful that we may have expected.

Currently we experience an advanced type of linear thinking and an eventual nonlinear enhancement may produce a much successful mental combination.

Second Cognitive Revolution

The threshold between these two cognitive revolutions is drawn by the language revolution archaeologically dated around 30,000 years ago.

In a preliminary stage, around 40,000 years ago, a dynamic language begun to replace the static language which existed for 30,000 years.

As it seems, the full dynamic language begun to manifest only 30,000-23,000 years ago.

Biologically, during the phase of the static language, it develops a gradual population of the optical (visual) tracts with acoustic signals.

However, the impact was relatively small, because the brain research did not show a significant evolution of the temporal lobe at this time.

As it seems, around 30,000 years ago, the hierarchical inversion that developed 70,000 years ago, producing holism, was reversed again, but this time by the development of the Frontoparietal cortex that attracted much of the additional pulses of neurogenesis and especially those from striatum neurogenesis.

This time the anatomical changes were significantly larger than at any previous time, while the result was radically distinct from holism. It arose **the linearization that radically changed the language and the conscious thinking**.

Biologically, it developed an increased latency in the long-distance tracts, which favored an increased population of acoustic signals, but also because of the speed reduction in the acoustic signal transmission more acoustic details became available and allowed a much better mental manipulation. The language became exclusively linear because the acoustic characteristics are linear.

The paramount effect in the new cognitive revolution was manifested in language development that increased ad-infinitum abstract outcomes, while turning them into symbolic manifestations. The second paramount effect of linearization was the quantification of qualities that occur in the same abstract realm. **Linearization, because of its fundamental superposition, generates an infinite multiplication of imaginary reality associated with its imaginary quantification**.

As mentioned, the **almost radical changes in the Homo sapiens' brain architecture created an outcome of radical nature that, in my opinion, led to a very distinct species that I called as the Speakingman or the Homo loquens**.

Conclusion

The changes in hierarchical-order, stimulated by bursts of supplementary neurogenesis, have occurred simultaneously to most planetary primates, contributing to short-range neural circuits development.

But we do not find a clear nonlinear behavior in these primates, which means, it occurs a significant inhibition of nonlinear outcomes in these animals probably caused by their primitive but powerful linearization.

R Robert Shapley in a study published online by Vision Res. 2009 Apr: 49(9):907-921 (doi: 10.1016/j.visres.2008.09.026) found:

"*Retinal pathways are surprisingly linear, and some neurons in the visual cortex also emulate linear sensory transducers. We conclude that retinal linearity depends on specialized ribbon synapses, while cortical linearity is the result of balanced excitatory and inhibitory synaptic interaction*. "

"*One reason that nonlinearity in neural information processing is the default is that neural communication is mainly through synaptic transmission, and most synapses are very nonlinear. The retina has very specialized synapses, the ribbon synapses and these specialized synapses appear to enable the retina to make use of linearity of signal transmission. The visual cortex must deal with the nonlinearity imposed by the spiking mechanism of spiking neurons that feed its visual input." "Cortical linearity is not simply the default result of convergence of excitatory inputs but rather requires extensive cortical computations."*

"*The simplicity and elegance of linear systems are created in the visual system, as a linearly filtered version of the visual world*."

"*Retina can operate in a linear manner by counting photons evoked events*."

Marcelo Bertalmio, Alex Gomez-Villa, Adrian Martin, Javier Vazquez-Corral, David Kane, and Jesus Malo published a study *Evidence for the intrinsically nonlinear nature of receptive fields in vision,* 2020, 10:16277 (doi.org/10.1038/s41598-020-73113-0) where the authors conclude:

"*Cortical neurons produce all-or-nothing action potentials instead of analog values, i.e. neurons are highly nonlinear, and* ***more recent studies have rejected the hypothesis that individual neurons can be modeled as a linear RF followed by an output nonlinearity, showing instead that each thin branch of the elaborated dendritic tree of neurons could be computing many different nonlinear combinations of its input***."

"*Retina has specialized cellular and molecular mechanisms that enable it to provide faithful, linear transformations of the visual image. V-1 cortex reconstructs and then selectively distorts the neural image.*"

The above-mentioned study indicates that **only a change in the brain computational power can lead to a nonlinear interpretation of the perceived visual inputs**

However, Dr. Rosenbaum hopes to find ways in which scientists can stimulate the human brain toward semi-balanced states, where nonlinear representations and processing are allowed.

In sum, those two cognitive revolutions humanity had experienced in the last 70,000 years seem to originate in a complexity of phenomena, but where the lead was taken by a switching from a balanced state to a semi-balanced state and then back to a balanced state.

Part Five

Some thoughts on Neurogenesis role on Human evolution, as a result of forcing of Solar Cycles and the Geomagnetic Excursions

A unique neurogenesis produced the modern brain.

Abstract

The aim of this paper is to document the eventuality that geomagnetic events of the last 450,000 years overlapped the hominins brain development.

The paper considers a certain pulsation of the geomagnetic and natural neural phenomena, which might lead to genetic and neurogenesis outcomes.

However, the changes in the hominin brain of the last 450,000 have been crucial in gradually generating the modern brain. On the other hand, the changes of the Homo sapiens brain were more distinct than those occurring in the Neanderthals, indicating a significant turning of the Homo sapiens neurogenesis.

Discussion

Joe H. Levine, Yihan Lin, and Michael B. Elowitz, in a study *Functional Roles of Pulsing in Genetic Circuits*, published in Science, 2013, Dec.6 (doi:10.1126/science.123999), indicate:

"*Many regulatory factors undergo continual, repetitive pulses of activation. Each of these pulses involves the coherent activation and deactivation of the regulator, through changes in its concentration, modification state, or localization, on time scales running from minutes to hours. The key regulators modulate pulse characteristics, such as frequencies and amplitude*."

"*In particular, we highlight the regulatory flexibility that comes from independently controlling the timing and amplitudes of pulses. We then discuss the circuit mechanisms that enable cells to generate and control pulsatile dynamics. Finally, we suggest additional ways, not yet discovered, in* ***which pulsatile regulation could potentially enhance cellular capabilities***."

"***Expression levels of target genes are then proportional to pulse frequency, and also to one another***. *This coordination function was demonstrated experimentally*."

Another study *Differential P1 arginine and lysine recognition in the prototypical protein convertase Kex2*, of Joshua L. Wheatley and Todd Holyoak, published by PNAS on April 17, 2007, explains:

"*The high-resolution crystal structure of kexin (Kex2) in complex with a peptidyl-chloromethyketone inhibitor containing a non-cognate lysine/arginine selectivity that defines the prohormone (proprotein) convertase family.* ***Studies demonstrate that the selectivity between lysine and arginine at P1 position arises at the acylation step***."

Therefore, the switching between lysine to arginine can be triggered by an overlapping exogeneous pulsation affecting the P1 position. Such an exogeneous interference is said to belong to an additional hydrogen-bonding present at P1 and absent in a lysine-containing substrate.

Possibly, other hydrogen-binding structures have been affected by overlapping pulses within the hominin's brains in the last 450,000nyears. **It and other contributing factors resulted in larger neural production in Homo sapiens compared to Neanderthals, while affecting a significantly bigger output in cerebellar area of Homo sapiens**.

However, the mentioned study concluded that "*the observed change in acylation rate between the substrates with P1 lysine or P1 arginine residue is the result of a more simplistic mechanism that is based on the observed binding of the P1 lysine residue*."

In my opinion, **the lysine's own pulsation can be overlapped or interfered with a direct or indirect exogeneous pulsation that would intermediate the switching toward arginine**.

This switching had occurred only in Homo sapiens, while the increased neurogenesis continued in all its brain structures, gradually and increasingly differentiating from Neanderthals.

"*Based on this structural information and the available kinetic data, it has been hypothesized that this S1 lid may close down only on a bound* ***P1 arginine and provide a large amount of energy to promote acylation step that would not be available in the presence of a P1 lysine***."

This approach is challenging in proliferating progenitor cells, like neurons, "*as dilution of cellular components during cell growth forces most protein concentrations to relax to their steady-state values over a time scale of about one cell cycle or faster.*"

In Science Daily, on September 8, 2022, was published the article "*Modern humans generate more brain neurons than Neanderthals*."

The article states:

"*Researchers from Max Plank Institute of Cell Biology and Genetics (MPI-CBG) in Dresden now show that the modern human variant of the protein TKTL1, which differs by one single amino acid from the Neanderthal variant, increases one type of brain progenitor cells, called basal glia, in the modern human brain. The researchers conclude that this single human-specific amino acid substitution in TKTL1 underlies a greater neuronal production in the developing frontal lobe of the neocortex in modern humans than Neanderthals*."

" *Specifically, in modern humans TKTL1 contains an arginine at the sequence position in question, whereas in Neanderthals TKTL1 it is the related amino acid lysine. Notably, the level of TKTL1 is highest in the progenitor cells of the frontal lobe*."

Another study of Martin Kuhlwilm, and Cedric Boeckx, titled "*Genetic differences between humans and other hominins contribute to the human condition*", published on April 11, 2018, in BioRxiv.

The authors said:

"*Human condition is rooted, at least in part, in the properties of the brain, and that these can be traced to changes in the genome on the modern human lineage. The phenotype in the population called "anatomically modern humans" emerged in Africa likely before the deepest divergence less than 100,000-200,000 years ago, although complex population structure may reach back up to 300,000years ago.*

It has been claimed that ***the brain of modern humans adopted a specific, apomorphic growth trajectory early in life that gave rise to the skull shape differences between modern humans and extinct branches of the genus Homo. This ontogenetic trajectory, termed "globularization phase", might have contributed to our singular cognitive abilities***."

"*Almost 1,000 putative protein-altering changes were found across most present-day humans. Three genes (CASC5, SPAG5, and KIF18A) have been singled out as involved in spindle pool assembly during mitosis. These have been highly relevant candidates for human-specific traits.*"

"*This suggests that the* ***cell cycle machinery might have been modified in a specific way in humans compared to other hominins***."

"*The number of neurons in the brain might be influenced by some of the changes in kinetochore-associated genes, and their organization and neuronal wiring clearly impose structural demands on the organization of the brain.*"

"*This might be linked to the* ***specific brain growth trajectory in modern humans and reflected in the morphology of the parietal and temporal lobes, as well as in the size of the cerebellum****. Archaic hominins likely had certain language-ability, and hybrids of modern and archaic humans must have survived in their communities. This, together with putative positive selection after the split from Neanderthals and regulatory changes affecting FOXP2 expression, could indicate modifications of a complex network in cognition or learning, possibly related to other brain-related, vocal tract or neural changes.*"

"*Apart from their cognitive functions, it has been suggested that changes involved in synaptic plasticity might be interpreted in* ***a context of neoteny, with the implication of delaying maturation in humans and a longer timeframe for brain development***."

Recent work in cell biology, after 2010, has unveiled the surprising role of pulsing in the circuitry dynamics of the living cells. Many genetic circuits actively and spontaneously generate **dynamic pulses, which temporarily organize the cellular functions**. Here, **"pulsing" is seen as a board spectrum of repetitive phenomena**. Each pulse generates the activation and deactivation of the regulator on time scales from minutes to hours.

I decided to present in brief this pulsing issue for the simple reason that **natural exogeneous pulsations will overlap the natural cellular dynamics, and sometimes, such an overlapping is responsible for well-defined genetic changes**.

The transcriptional program that controls the gene expression is flexible, constantly adapting to various exogeneous perturbations. Different classes of **temporal patterns are influenced by external stimulus, such as transient environmental stimuli**.

As it appears, the most significant changes in hominins brain occurred only in the last 450,000 years, starting with cerebellar area development, which was enhanced in the members of the Last Clade, and continued to significantly but independently evolved in Homo sapiens compared to Neanderthals.

Recent research, as mentioned above, now indicates another major functional distinction between Homo sapiens and Neanderthals that refers to the switching from amino acid lysine to amino acid arginine.

As the research found, this switching produced an enhanced neuronal output in the development of the frontal lobe.

L-arginine consists of the most nitric oxide supplements, which may have been provided by the prehistoric diets, which took the form of a rudimentary L-citrulline. Consumption of such citric food in the African environments may have contrasted to its availability for Neanderthals in the Eurasian environments. Arginine generates a reduction of the metabolic oxygen consumption that might allow more oxygen provided to the brain.

However, the dietary enhancement provided by arginine-bearing natural foods is strongly debated.

The enhanced neuronal output of the neurogenesis certainly generated a unique development of the parietal lobes in Homo sapiens that is not found in Neanderthals. Such development occurred starting some 100,000 years ago, but it became overwhelming in the era 70,000 to 23,000 years ago. This last process led to the globularization of the human skull that is synonymous to the setup of the modern brain that becomes too, a language-ready processor.

Carlos Wilson, Ernesto Munoz-Palma, and Christian Gonzales-Billault published the study *From birth to death: A role for reactive oxygen species in neuronal development*, published Science Direct, Vol 80, August 2018, pages 43-49 (doi: 10.1016/j.semcdb.2017.09.012).

The authors stated:

"*We discuss* ***how ROS and redox state contribute to neurogenesis, polarization, and maturation of neurons, providing a context for the spatio-temporal conditions in which ROS modulate neural fate, discriminating between "oxidative distress", and "oxidative eustress.***"

"*This review focuses on neuronal effect of ROS mediated by the NOX complex, discussing their contribution to differentiation, development and regeneration of neurons of both the central and peripheral nervous system.*"

"*Accumulated evidence suggests that ROS should be considered as second messengers involved in numerous signaling pathways in health and disease. Undoubtedly, abnormally high, and dysregulated ROS production leads to oxidative distress and cell death, but* ***the regulation of ROS levels in response to cellular demand is critical for normal cell behaviors****.*

Altogether, ROS supports the oxidative power of the eukaryotic cell, promoting important physiological functions, ranging from innate immune response to neuronal development. Of note, ***molecular (N2)****, the most abundant gas in troposphere,* ***also drives the synthesis of pro-oxidative molecules, collectively named Nitrogen Reactive Species (NRS)****, as N2 represents almost 78% of the air we breathe.*"

Other studies have studied the effect on hypoxia/re-oxygenation in neuropheres cultures. They found that **the regeneration production is dependent on ROS production that enhances the number of neurons 1.4-fold. There is a clear indication that NSCs resist any ROS related damages.**

Considering that the hominin brains have been influenced by some pulsating environmental conditions, like the atmospheric C14 isotopes, while overlapping natural neuronal pulsations, it is important to identify such **exogeneous occurrences**.

Around 450,000years ago it occurred a change in the glaciation pattern, making the differentiation between glacial and interglacial stages more thermally distinct. In other words, the glacial phase turned cooler.

In the meantime, **the occurrence of geomagnetic excursions, where some implied temporal geomagnetic polar reversals, became more frequent, causing increased intervals of weak exposure to the cosmogenic radiation**.

For example, the five known excursions during the era 480,000 to 300,000 years ago cumulated a combined excursional mode of 60,000 years. However, from 300,000 to 150,000 years ago occurred another four excursions with a combined cosmogenic exposure of 40,000 years.

Cosmogenic exposure can be characterized by two components. One is the amount of the harmful UV radiation that penetrates through the ozone holes generated by the excursion. Not much of such an effect has been documented for the last 500,000 years, because it will relate to mass extinctions. Only a few extinctions can be documented.

The only significant while well-documented effect of the penetrating cosmogenic radiation to the ground manifests by an increase of the atmospheric concentration of C14 isotopes. Such concentration can double the regular atmospheric level of this component.

The C14 isotopes are processed by the photosynthesis of plants and thus they enter the food chain, affecting all biota.

I wrote about the role of C14 isotopes in influencing human neurogenesis in several articles, which are uploaded on my site on academia.edu.

Here, I would like to recall the **pulsative effect of the C14 isotopes, which is constantly or partially absorbed or diluted by various environmental sinks. Such a process implies a repeated pulsation on the scales of months, weeks, or even days**.

Therefore, the overlapping might have a significant effect, leading to genetic changes and/or changes in genetic expression, those which regulate the neurogenesis and the components of the brain architecture.

Returning to the geomagnetic events, the epoch 125,000 to 5,000 years ago contains the most numerous geomagnetic excursions, including four temporary reversals. In this era occurred some ten to eleven geomagnetic excursions.

The Blake and the post-Blake events (125,000 to 90,000 years ago) had almost 6.5 ka years of switched polarity divided into around three occurrences. The longest was 4 ka. **This era dates for the first archaeological finding of the "modern" Homo sapiens**.

Another temporary polarity reversal occurred 41 ka during the Laschamp Event. Archaeologically **this epoch dates the first "modern brain" as a language-ready brain and an associated language revolution. It is the epoch when the Frontoparietal Synthesis begins to manifest**.

A hypothesis on Human evolution, as the result of forcing by Solar Cycles and Geomagnetic Excursions

Christopher J. Camprisano (institute of Human origins, School of Human Evolution & Social Change, Arizona State University) 2012 Nature Education, published in Knowledge Project a study *Milankovitch Cycles, Paleo climate Change, and Hominin Evolution.* The author said: "*It seems intuitive that large-scale shifts and short-term variability in paleoclimate altered regional hominin habitats and resource availability that ultimately led to selection pressure on our fossil ancestors*."

"*Variability Selection Hypothesis (Potts 1998) proposes that the wide variability in adaptive settings over time ultimately favored complex adaptations that were responsive to novel conditions (i.e., the evolution of adaptability)*."

"***There is no single "magic bullet" that is responsible for the multitude of anatomical and behavioral changes in the hominin record***..."

"*All hypotheses that propose causal links between paleoclimatic change and hominin evolution must ultimately reconcile global patterns with local responses and extend far beyond a general temporal correlation between environmental change and evolutionary event.*"

I would like to say that everything is nonlinear and complex around us, and thus, no linear bullet can provide an answer. Even then, some multidisciplinary approaches might get close to an answer.

In the meantime, all these cycles are wave phenomena, which are subject to wave composition and interference. There is reasonable to think that such waves composition would vary in terms of Sun own energetic resources and particular mechanism that we do not know, yet. Hence, it may occur various combinations, producing distinct outcomes, which all show a time variability.

However, the same variability impacts the Earth's thermal output, while it might combine with the Sun variability. Not to mention that cosmogenic radiation is another wave variable, which will tend to interfere with the other two mentioned exogeneous factors.

The above-mentioned author just mentioned in his study that paleo climate, (that is largely associate with the earth's own magnetic field and other forcing factors, like our sun) can affect Hominin evolution of the Last Clade.

Why the Last Clade (or the last 450,000 years)? It is so, because significant changes have occurred within the Sun radiative output, along with changes in the Earth's thermal variability that is associated with the geomagnetic factor.

And then, we have documented a change in the architecture of brains of the members of the Last Clade that occurred only in the last 450,000 years. Such cerebral changes separated the Last Clade hominins from monkeys.

Interestingly, during the same era of the last 450,000 years, some cerebral changes occurred in the **erectus's brain**, in terms of an increased brain volume and an associated cognition. Even then, these changes were not enough to pass the threshold away from monkeys. Among other things, erectus became less than a million years ago the first bi-pedal to run a big deal on the ground, and to adapt fast to changed environments. He became extinct only 100,000 years ago in Indonesia.

Hence, the erectus brain encountered 350,000 years of evolution due to similar forcing that occurred all over the planet, which have been the same forcing that made the species of the Last Clade differ from monkeys. Due to last half a million years forcing, the erectus brain evolved in China from 800 cubic centimeters to 1,230 cubic centimeters, almost on par with the Last Clade hominins. Even then, erectus does not show the same bulging of the occipital skull like in Neanderthals and Early Homo sapiens. It means, cerebellar area of erectus did not evolve significantly. The same case occurred in his frontal, parietal, and temporal lobes. But his brain grew significantly in volume.

As another example, I will mention the Early Homo sapiens, Neanderthals, Denisovans, and Homo heidelberensis, all members of the Last Clade. In the first phase, starting 450,000 years ago, in all of them have occurred a unique development in the cerebellar area. This aspect enormously improved their cognition, in the manner that they become distinct from monkeys.

However, Homo heidelberensis became extinct first. Denisovans survived until cca. 50,000 years ago, in South Spain. Neanderthals evolved the most, while lacking a coherent language, a significant frontoparietal development, and having a slower cerebellar development compared with sapiens.

I would say that each species of the Last Clade evolved differently, and eventually some distinctions were created between groups or even individuals. It was about different genetic expressions caused by differently evolving genes. The differentiation has manifested at the behavioral and cultural level, which cannot be detected by the fossil skull analysis. The Neanderthals are known to possess quite a rich cultural life, while lacking many other genetic skills present in sapiens.

This case indicates that hominin of the Last Clade, and erectus brains were subject to common exogeneous forcing factors, but their individual cerebral evolution resulted in distinct ways and different life-styles mentalities and strategies.

There is no doubt that all brains of the Last Clade members were vastly affected by such exogeneous factors.

However, any brain development highly depends on its own neurogenesis. Here, must have been factors which contributed variably to each brain neurogenesis. Homogenization was assured by species matting, and during socialization activities and practices.

Most of the brain changes occur in the prenatal and immediate postnatal stages. But the stem and progenitor cells produced by neurogenesis in the embryonal phase encounter differentiation only later in childhood and teenager years.

Some genes serve neuronal proliferation, allowing a neuron to be plastic, meaning that it can adapt to multiple tasks. Other genes support a further process of specialization, during which a part of the neurons becomes strictly adapted to particular tasks for which they develop abilities named skills.

Thus, the **exogenic forcing factors must be acting mostly in the prenatal phase when the genes will configure the later neuronal network**. **Inheritability must play a role, too**. The skills are acquired and formed much later in life.

* * *

I would like to quote an article wrote by Bjarne Lorenzen (Denmark) who published his study *Earth's Magnetic field-The Key to Global Warming* in the Journal of Geoscience and Environmental Protection, 2019,7,2538 (http://www.scrirp.org/journal/gep, ISSN Online:2327-4344, ISSN Print: 2327-4336).

The author explains.

"*One possible cause and effect connection may (in theory) be that inner heat in the Earth's crust can loosen frozen, ferromagnetic structures, thereby drive the Earth's ferromagnetic, magnetic field to restructure and be organized from periodically being a chaotic, magnetic field being well-structured ferromagnetic field. The connection between magnetism and thermal impact is already known.*

Henrik Svensmark's (and team) theory says that variations in the cosmic radiation reaching the Earth depends on the strength of the Sun's magnetic field and that this radiation contributes to creating aerosols, thereby variations in the cloud formation. *Solar storms contribute to the temporary strengthening of the Earth's magnetic field.*"

"The new theory indicates **that the MF (Magnetic Field) of the Earth is simply a ferromagnetic field, solely generated in the Earth's crust**."

"Before the "inner geodynamo theory" took over, the scientific community accepted that the MF was a ferromagnetic field created by the materiel in the crust."

"*The Earth periodically goes through periods that are dominated by "magnetic chaos." During these periods of fragmented magnetic activity, two (or even four) Magnetic North and South Poles exist, and the magnetic strength of the global field is very weak, or around 10 percent from normal.*"

"***The new theory calls this phenomenon: "ferro-magnetic integration***."

The author explains how it works the new theory:

"*Deep inside the crust of Earth (where temperature is higher than on the surface (but still bellow Curie point), there are optimal conditions for **the process of magnetic integration**. The "magnetic resistance" is lower in these warmer areas due to disruptive, "frozen" magnetic structures and is significantly reduced.*"

"*The two great norther continents have two strong magnetic areas: one above northern Siberia and the other one above North America. During weak magnetic periods, disintegration of the global FMF of the Earth will first reduce the range. Finally, Eurasia and South/North Americas will have their own north and south poles.*

*This is what is happening right now like continual global disintegration (that started 2700 years ago during Stern-Etrussia excursion). This is the main cause for **some magnetic disentanglements followed by new entanglements.***"

"*The magnetic weakness that we currently see in the Atlantic Ocean (called The South Atlantic Anomaly) is in reality the first serious sign of a period when the global MF has begun the process of separation into two parts.*"

"*In the Arctic Ocean, we find one of the biggest magnetic anomalies on Earth. The Arctic Ocean is the natural place where the two big continental ferromagnetic fields are gathered into one global field.*"

"During such periods, the MNP (Magnetic North Pole) is entangling many of the Earth's magnetic power lines by pointing them into ***increasingly chaotic magnetic directions****, while it* ***causes the disentangling of those two continents****."*

"The faster and wider the NMP travels over the "cool" chaotic magnetic lines on its way, ***it causes the reduction of the strength of the global MF (Magnetic Field)****."*

"The strength and size of the continental magnetic anomalies are partially caused by materials that can be magnetized in the Earth's crust often plays a major part."

"It is very likely that the magnetic field of Mars once was a much stronger magnetic field. The ***loss of thermal heat in the crust of mars caused the magnetic structure to freeze in chaotic directions as the crust of the planet was cooling off****. What is seen on Mars is that some magnetic alliances (in the crust) remain, but they are very small and probably completely "frozen" and unable to gain any significant size and strength, even locally."*

"***Periods of a weak and chaotic MF are always followed by periods when, once again, a global FMF is created***."

"If frozen flux in the Atlantic Ocean begins to dominate-that area will lose its ability to magnetically integrate with global MF. This will cause accelerating magnetic disintegration, the magnetic chaos begins to rule, and repeated magnetic pole reversals can easy take place."

The reason I quoted so extensively the above paper is because it brings out planetary mechanics in the realm of The Science of Complexity.

As the Science of Complexity describes, the elements, amounting to a complexity, need energy for their relationships to occur. When such energy turns to low, the parts tend to dissociate from the whole, **producing a collapse of complexity**. **Such parts become chaotically oriented and might search for smaller alliances**. An insertion of exogeneous energy would stimulate again the complexification process, turning small alliance into larger ones, which form new complexities.

It sounds like the ferromagnetic process described above by Bjarne Lorenzen.

(However, the astronomical description of Milankovitch's cycles indicates **three basic components**, **eccentricity**, **obliquity**, and **precession**, which have the respective lengths of 100 ka, 41 ka, and 19-23 ka. Beyond that, we know that the 11 years sun cycle is in fact a 22,000 years cycle, where the maximum and minima occur twice. All these come in support of the mentioned writer of the article in question.)"

The **author's depiction of fragmentation followed by fragmentation-reorganization (integration), allegedly occurring in the Earth' crust, represents some <u>complexification/decomplexification processes, affecting the ferromagnetic material of the crust</u>**.

As the author explained, the Earth's magnetic field gets strength when the Sun sends to it more radiation. If the Sun obeys certain thermonuclear cycles, the Sun could maneuver the Earth's energy during such solar thermonuclear cycles. There are other astronomic factors, which may contribute to refreshing Sun energy, like when the solar system crosses the cosmic clouds of stellar dust, which populate some galactic regions that the Earth may randomly traverse.

It might show that our planetary crust encounters complexification and complexification collapses, as frequent processes stimulated by the Sun. **Such processes are vividly pictured during the geomagnetic excursions**, when the collapse of the crustal ferromagnetic field takes place. It is followed by chaotic magnetic directions, and this chaos destroys the wholeness of that field.

Hence, the frequency of the field drops to almost zero (10% remaining magnetization generated by small ferromagnetic alliances is almost zero) and allows the cosmogenic radiation, the galactic one that is more energetic than the Sun's one, penetrate to the ground in the form of nuclides or secondary radiation, like the C14 isotopes.

The only thing the penetrating cosmogenic radiation generates is a high atmospheric concentration of C14 isotopes, which isotopes are absorbed by all biota, including hominins and humans. **No damaging ionizing**

cosmogenic radiation can cross directly to the ground, except the UV radiation that would penetrate trough some ozone holes, if they can be formed in the stratosphere.

Usual ozone holes are created near Antarctica, are over the southern oceans and of no harm to humans. Some geophysical theorists had speculated that, especially during the temporal pole reversals, such ozone holes would multiply all over the stratosphere. But this is still a theory, because if it is correct, it means that during every excursion with polar reversal much of the biota will turn extinct. We have at least five to six-if Mono Lake complex excursion qualifies for it- in the last 115,000 years.

At the end of previous interglacial, named Eemian (130-115 KA), occurred the **Blake Excursion**, which had a 650 years pole reversal developed around 119 ka, and a second reversal took place for 4,500 years centered around 112 ka. During the post-Blake event (98-96 ka), took place another two excursions with sharp pole reversals.

The fifth excursion, the Laschamp excursion took place 41 ka, and the pole reversal lasted from three hundred to one hundred years.

By contrast, regular excursions without pole reversal have been very numerous. **Albuquerque Excursion developed 146, 000 years ago** without a polar reversal.

Eventually, the Albuquerque excursion served neurogenesis with an additional but needed "drop" that **helped 130,000 years ago**, at the beginning of the Eemian Interglacial, **various group of sapiens to advance into "moderns"**. It resulted in a species that showed distinct features, when compared to their previous stage called Ancient or Early Homo sapiens (AHS or EHS); but most importantly, now they were very distinct from Neanderthals.

For example, the **Norwegian-Greenland Sea Excursion** is in general documented at 66 ka and a second leg at 55 ka. But in Northwest China, and in Bangladesh the specialist found it for a duration of 10,000 years. **In China was named Gaotai** and lasted from around 82,8 ka to 72,4 ka A similarly duration and dating was found for this event in Bangladesh.

It was followed by **Laschamp event** 41 ka that developed a pole reversal. The **underlying excursion process lasted 5 ka, from 43 ka to 38 ka and the surface field expression lasted 2 ka from 42 ka to 40 ka**, leading to more than one excursion between 36 ka and 30 ka.

This excursion was followed by **Mono Lake** at 33 ka, but in fact it had many legs (34 ka to 30 ka) centered on 31 ka, **or it has been a series of excursions or a small group of several excursions with aborted and successful reversals**.

In China, Mono Lake Excursion was found 28.4 ka to 25.8 ka. It shows how widespread was the defragmentation of the geomagnetic field, which became prone to easy occurring reversals, where some were so short that cannot be correctly traced on geological samples.

The specialist in geophysics likes to say that between 50-26 Ka was an almost continuous low frequency field, which depicts several major phases of defragmentation followed by rapid but short-term reintegration.

Just to confirm the above idea of continual changes in the planetary MF from 50 to 25 ka, almost in the same time interval was defined another excursion named **Lake Mungo**. It basically overlapped Mono Lake because it is described as from 31 ka to 28 ka. Allegedly, it was generated a second excursion at 26 Ka that indicates a possible aborted reversal.

This excursion preceded **Last Glacial Maximum** (LGM 25 ka to 19 ka).

One can see that such instability of the MF (Magnetic Field) for 25,000 years led to the **Last Glacial Maximum** that started 25,000 years ago and indicated a large loss of crustal thermal energy, inflicting a huge chaotic movement of the ferromagnetic components.

A similar situation occurred during the Blake and post-Blake events (119 to 96 ka), when four reversals occurred, totaling over five thousands of years of reversed polarity, and which events caused a **Glacial Maximum** (70 ka) that, even being shorter than the LGM, it **was in many respects more severe**.

Another expression of planetary complexity comes from the MF instability of the post-Blake events associated with the Norwegian

Greenland Sea event that evolved distinctly on the Chinese-Bangladeshian axis and stimulated 74,000 years ago the **Mount Toba volcanic catastrophic eruption**. It was followed by one to two years of volcanic cooling from Indonesia to Tropical and Equatorial Africa. The severity of the effect in Africa is still largely debated, being infirmated by some contradicting evidence.

In the meantime, the excursions, occurring around 30 ka could be associated with Nemrut yanardagi strong volcanic eruption in Anatolia (Turkey) at 29.7 ka and Aira Caldera, Kyushu (30 ka, Japan), and Campi Flegrei (near Napoli, Italy) also known as the Phlagraean Fields (29.1 ka) were so powerful that caused local animal extinctions in Southern and Eastern Europe.

The volcanic excursions intervene when the atmospheric thermal status turns high as a result of solar radiation and ground reflection of it. The volcanic ashes increase the cloudiness by flying ashes for one year or more. It generates, depending on the size of the eruption, a general or local cooling on the ground.

Returning to the analysis of the excursions, it should be recalled the fact that during the excursions, the geomagnetic field is weak, meaning that it is dominated by chaotic orientations of the ferromagnetic crustal material.

At around 17 ka was **Hilina Pali Excursion** that ended the Last Glacial Maximum and was found in various location at different age, like at 20 ka, 19-17 ka, or 17.3 to 16.9 ka (or 20 Ka to 17 ka).

It was followed by the **Gothenburg Excursion** between 12,494 and 13,081 years ago. An interfering cooling event, of Younger Dryas (12.9-11.6), overlapped the last part of this excursion, probably being generated by it. It postulates that a phase of rapid warming that started 18 ka, did not have enough energy to support its phase of complexification, following the collapsed complexity during the LGM (last Glacial Maximum).

An unnamed excursion was determined around **10,000 years ago** in Ecuador and Argentina.

The **Solovki Excursion** 7,500 to 4.500 years ago has a less important atmospheric concentration of C14 isotopes, but it still reached 40% above normal level around 3,500-4,000 years ago.

However, the excursion reached a minimum of the geomagnetic frequency/strength (maximum cooling, too) around 6,000 to 5,000 years ago, when a severe cooling event took place especially at higher altitude. It was a surprise for the Ancients seeing the glaciers coming back all over the mountains.

The last excursion was the **Sterno-Atrussia Excursion** that started 2,814-2,813 years ago and lasted for 200-300 years.

I would like to mention that during the LGM the planetary land surface was covered 30 percent by ice sheets and snow.

I would like to mention that geophysical recent studies identified samples with stabile and instable records with normal, intermediate, and reversed directions, which occurred at different times during the last 11-10 Ka in Southern Patagonia and Chile. Also, in Chubut province in Central Patagonia were identified reverse directions at 2 ka.

In Southern Chilean Patagonia, Mylodovan cave yielded several reverse directions between 10 ka and 5.5 ka. All such data leads to the proposal that a Geomagnetic Excursion occurred in Southern South America, locally named Mylodovan Excursion, which implies a short reversal, too.

According to Nami (2012), during the last 11-10 Ka, the normal polarity of Earth's magnetic field has been interrupted by several short-lived reverse polarity events.

Saul authored a paper "*Geomagnetic Excursions Interglacial Termination Abrupt Climate Change*," published in Earth Sciences, March 5, 2010.

The author explains:

"*The periodicity of the forcing event that caused the geomagnetic excursion is around 12 Ka. The same forcing event, at a smaller magnitude, causes geomagnetic jerks which are secular geomagnetic field changes in regions, where have stronger normal or reversed polarity.*"

"**Archeomagnetic "jerks" are abrupt secular changes to the geomagnetic field intensity with periodicity of roughly 400 years**."

"It appears archeomagnetic jerks, geomagnetic field excursions, and geomagnetic reversals are being forced by significant cyclic solar events.

* * *

The extinction eventuality cannot be documented in the existing animal fossils. Some extinctions took place but cannot be placed near the scale implied by a large multitude of stratospheric ozone holes.

However, the length of each geomagnetic excursion depends on the time needed to accomplish the Earth energy recharging needs. The excursions can be thousands of years long, or only a hundred or a few.

* * *

However, **the mentioned author theory, in my opinion, does not completely eliminate the "the inner geodynamo theory", but only it changes the entire picture to accommodate an assembled together work with the "ferromagnetic approach**."

Another contribution of the mentioned author regards the solar influence on our planet magnetic field, defining de novo the mechanism of the glacial/interglacial cycles.

"*It is known that the Earth's MF is temporarily "raised" with the contribution of the solar storms.*

"***Long-term coincidence between climatic temperatures of the Earth on one hand and the global strength of the MF of the Earth on the other is well-known and an accepted fact***."

"*The coinciding data clearly indicate that the cause of ice ages is almost completely misunderstood and the Milankovich cycles are far from enough to fully understand the cause-effect process*."

"The periods of solar activity are not always brief (which means not only 11 years or periods a few hundred years) but also periodic solar activity cycles in the last 800,000 years have lasted 100,000 years; before that they were 41,000 years with an equally displace glaciation/interglaciation epochs. Five million years ago the glacial eras were less than 15,000 years.

The author hypothesis becomes stronger when one considers the theoretical overlapping of various solar cycles. It implies that the sun itself is the subject of such cycles. It means, we currently experience the 11 years cycle that is overlapped by a larger cycle (like the one identified for 20,000 years, and the one found to last 41,000 years), and where a much larger one of 100,000 years circumscribes all others.

It might be the case that the sun developed various cycles, which exist simultaneously, and/or it encountered a process where some of its cycles become predominant during various geological times.

In sum, I would like to say that in the last five million years, sun developed a dominant cyclicity of 15,000-10,000 years cyclicity, which around less than three million years turned to a dominant cyclicity of 41,000 years, where opposite geomagnetic intensities were equally displaced.

And then, around 1,500,000 to 500,000 years ago, the sun cyclicity changed again, causing a dominant low frequency for 90,000 years and a high frequency for 10,000-15,000 years. During this last cycling change, the glacial era became from not a big difference between glacial and interglacial, toward more intense difference in temperature.

Another change was toward a warmer era with peak temperatures of only 0.5 to 2.0 Centigrade. Such episodes started 720,000 years ago and were named **Chibanian Era** that extends to 130,000 years ago).

The mentioned warmer episodes, which occurred before the Esterian, or before 435,000 years ago, ended 425,000 years ago, when the **Holsteinian interglacial** began. It was the longest interglacial (425,000 to 395,000 years ago) on record (30 ka). In fact, this was a turning point in planetary glaciation episodes.

During 140-ka long **Elsterian Glaciation** (570,000 years ago to 425,000 years ago), were three warming episodes, where two of them were under zero Centigrade, and a very short one was 1 Centigrade.

From 395,000 years ago, the cold episode became colder, and the warmer ones, warmer.

During the **Saalian Glaciation** the cold episodes extended to 90,000 years, while the warmer ones diminished to 15,000 to 10,000 years.

The last warm era before the last glacial, named **Eemian**, reached the highest temperature on the 720,000 years scale with a peak of almost 5 Centigrade.

The **Saalian Glacial** era went from 330,000 to 240,000 years ago, defining a 90-ka episode, while after a long but relative warm episode, it produced significant low temperatures at the ending of this sequence.

Between Saalian Glaciation and Eemian Interglacial occurred two short interglacials and two glacial eras.

However, let's see the meaning of the excursional mode of the geomagnetic field as it might have generated significant turning moments in the evolution of the brain architecture in Homo sapiens in the last 100,000 years, and when such changes have been anatomically recognized by the major anthropological studies.

Here, it is the Blake and post-Blake effects (112,000 to 95,000 years ago) that caused significant increases in C14 isotopes concentration in the atmosphere. This era corresponds with an initial startup of the globularization of the huma skull that in fact concerns the evolution of the precuneus as the main anatomical expansion. In the meantime, the precuneus unique expansion in sapiens seems to have started only around 70,000 years ago. Thus, some other neuronal changes occurred between 100, 000 and 70,000 years ago, probably in the frontal, occipital and temporal lobes, increasing cognition and plasticity of the brain. Essentially plasticity, caused by significant increases in neurogenesis, accounts for generating new abilities in human adaptation, and it represents the main cause that led to the out-of-Africa migration defined at 55,000 years ago.

As it appears, the Norwegian-Greenland Sea Excursion with alternating variation developed between 82,000 and 64,500 years ago had been an additional triggering factor in increased plasticity, but also it appeared to have initiated the precuneus expansion since 70,000 years ago.

This moment, dated 70,000 years ago corresponds with **the first documented art expression at Blombos Cave in South Africa. It indicates a mental change toward generating abstract thinking**.

As explained above, the low era of geomagnetic field strength, from 50,000 to 25,000 years ago is the main candidate in reshaping the human brain architecture. Among the products generated in this era, I would mention the transition from a static while primitive language into a dynamic one by the insertion of verbs and adverbs, which occurred prior to 30,000 years ago. This new approach is directly connected with a precuneus capable of making mental visual constructs, which capability is defined by the Frontoparietal Synthesis development that maturated only 30,000 years ago.

In sum, 30,000 years ago the brain became language-ready, as it is archaeologically proven as a language revolution dated 30,000 years ago. Simultaneously, the precuneus development generated the first skills for mental composition that started with the language, and it spread in mentally composing, analyzing, and synthesizing from abstracts to physical constructs.

Here comes another theory that relates skull globularization generated by the precuneus with processes of self-domestication and grammaticalization of the mind (see Antonio Benitez-Burraco, Constantina Theofanopoulou, and Cedric Boeckx study Globularization and Domestication, published in 2016, doi: 10.1007/s11245-016-9399-7).

"*The link between globularization and neural crest function (and between language-readiness and language change via domestication) is a strong one, and results in a distinctive modification of the cognitive phenotype of ancient hominins.*

The globularization of the brain and braincase sems to be related to emergence of other traits that we believe important for modern language, including the refinement of the neural substrate for "externalization (via ROBO/SLIT/FOXP2 interactomes) (see Boeckx and Benitez-Burraco 2014 for details) or new inter- and interhemispheric connections underlaying modern cognition (via the SHH-GLI pathway).

A globularized brain and brain case may also be linked to the developmental/genetic context for a domestic phenotype.

Put another way, globularization may be part of Deacon's (2009) discussion of "relaxation of selection" leading to a niche favoring cultural selection.

This is not to say that globularization and self-domestication are one and the same. It is well established that globularization led to a critical change at the level of the parietal lobe and associated parietal bone (Bruner 2004, 2010; Bruner et al. 2015).The posterior braincase changes were driving globularization, as Bruner contends."

However, as it is an archaeological fact, **not significantly globularized brain cases were found younger than 35,000 years ago**.

Let's have a summation of **the main Anthropocene changes in the last 15,000 years** that is the era when humans begun to significantly evolve at the end of the Last Ice Age (LIA).

Rivers are central to debate this subject because the river effects were more varied and intense than commonly has been recognized.

Several studies identify five stages as follow:

-**minimal effects** before 15,000 years ago with the use of fire and gathering of plants and aquatic resources.

-**minor effects** from increased cultivation after about 15,000 years ago, with plant and animal domestication after about 10,700 years ago

-**agricultural era** after about 9,800 years ago with widespread fire use, the first dams and irrigation, and mud-brick manufacture

-**irrigation era** from about 6,500 years ago with large-scale irrigation, major cities, the first large dam, urban water supplies, expanded ground water use, river fleets, and alluvial mining.

-**engineering era** with embankments, dams, and watermills after 3,000 years ago.

-**technological era** after 1800 CE.

However, the first evidence on incipient agricultural development was identified on the shore on the Lake of Galilee (Israel) and dated 23,000 years ago during the full blow of the LGM; at the same site some temporary settlements were also found.

An incipient use of fire for clearing the land for various uses, including hunting, was identified on the shores of Lake Malawi (East Africa) around 30,000 years ago.

The proofs about the use of the Frontoparietal Synthesis are all dated around 30,000-23,000 years ago. They include funeral evidence, bow making, advanced stone tools, animal domestication (dogs in China 15,000 years ago), and first temporary huts of temporary or seasonal settlements.

Part Six

The Future of the Speakingman

The Future of the Speakingman

Abstract

The aim of this paper is to discuss the trends manifesting in the modern brain and their evolutionary impact. I start with the diminishing of the brain volume, its causes, and implications. It shows the role played by the externalization of knowledge produced within the development of social system, the social distribution of cognition along systems of storage and sharing of information.

Other causes contributing to cerebral volume diminishing are considered, too.

The next current trend is represented by the unnatural selection and functional cerebral alterations, both being caused by the revolution produced by digital technology and media.

Further on, the discussion analyzes the effect caused by social swarming. However, here the discussion trays to unveil the physical mechanism that generates swarming.

A significant development of oculomotor system indicates changes in the speed of input processing because of significant increase in the saccade latency. This paper trays to demonstrate, from the analysis of text reading, that this mentioned evolution made humans see more details of the environment, increasing the discovery capability.

It follows another discussion on other aspects involved in the swarming mechanism and how this mechanism could be stimulated anew in the human brain.

The paper ends with quotes from the leading scientists about future social development.

Discussion

Diminishing in our brain volume: implications and consequences

Jeremy M. DeSilva, James F. A. Traniello, Alexander G. Claxton, and Luke D. Fannin published in October 2021 in Frontiers in Ecology and Evolution a study *When and Why Did Human Brains Decreased in Size? A New Change-Point Analysis and Insights From Brain Evolution in Ants*.

The authors explain: "*We find that hominin brains experienced positive rate changes at 2.1 and 1.5 million years ago, coincident with the early evolution of Homo and technological innovations evident in the archaeological record. But we also find that **human brain size reduction was surprisingly recent, occurring in the last 3,000 years**. Our dating does not support hypotheses concerning brain size reduction as a by-product of body size reduction, a result of a shift to an agricultural diet, or a consequence of self-domestication.*" "*We suggest our analysis supports the hypothesis that **the recent decrease in brain size may instead result from the externalization of knowledge and advantages of group-level decision-making due in part to the advent of social systems of distributed cognition and the storage and sharing of information**. Humans live in social groups in which multiple brains contribute to the emergence of collective intelligence.*"

"*In primates, greater cognitive challenges associated with forming bounded social groups in large societies, among other influences, appear to have selected for increased brain size.*"

"*Across diverse species that vary in social organization, **cognitive demands on individuals may be lower in societies in which group decision-making is more efficacious than individual decision-making. Collective intelligence may reduce brain size in humans and ants**. Therefore, the size of groups and society-level intelligence may affect* *behavioral performance and cognitive loads and increase or reduce brain size, depending on context.*"

Regarding the reduction in humans, "*some (*researchers*) have placed its origin in the late Pleistocene or 35 ka and others in the more recent Holocene or 10 ka.*"

"*Encephalization results from a trade-off in allocated resources from one tissue (the brain) to another (the gut) and was made possible by shifts to a higher quality diet in Homo.*

However, ***with growing complexity in social life, perhaps associated with increased group size, brain expansion occurred. Resources sharing within groups and allocate may have provided the energy surplus needed to support the increased energetic cost of a larger brain****.*

One possibility is that it is associated with a corresponding decrease in body mass. The early expansion of the brain in Homo has been explained as an increase in absolute body size. The actual reduction (Holocene) is more than 5 x greater, suggesting that body size alone cannot entirely explain the decrease in brain volume. Ruff et al., (1997) notes that the decrease in body size in late Pleistocene humans began around 50 ka, whereas brain reduction appears to be a Holocene phenomenon.

The brain size reduction in Holocene humans' parallels that of domesticated animals*, suggesting that humans have self-domesticated by deliberately removing highly aggressive individuals from breeding populations, leading to a reduction in intra-population (but not inter-population) aggression.* ***Brain reduction in this case would be a by-product (or cause) of docility****, a phenomenon documented recently in domesticated cattle (Balcarel et al., 2021). Grove (1999) suggests that* ***domesticated dogs have sufficiently co-evolved with humans to have symbiotically become our external senses, thereby decreasing our reliance on brain centers that process sensory information****.*

Human self-domestication is argued to have occurred at the onset of species-300,000 years ago (Wrangham 2019) or coincident with the evolution of what some have called "behavioral modernity -80,000 years ago (Hare and Woods 2020."

"***Reduction in brain size may not compromise cognitive performance if intelligence is an attribute of the society rather than the individual. Collective intelligence may reduce demands for neural architecture required to support some aspects of individual cognitive capabilities.*** *Using a multidisciplinary modeling approach, Resendiz-Benhumea et al., (2021) indeed demonstrated that* ***agents with relatively small brains can through social interaction achieve a level of behavioral performance capable to those of larger-brained but solitary agents.***"

"***We suggest that group cognition lowered the demands for neural architectures required to support some aspects of individual intelligence and decision-making (Bailey and Geary, 2009). This effect may have become more pronounced with the advent of writing ca. 5,000 years ago.*** *(Schmandt-Besserat, 2010).*"

"***A rise in sociocultural complexity was not due to particular individuals becoming more intelligent and culturally skilled, but because of the emergence of collective intelligence resulting from growing population of interconnected humans and interacting human groups.***" "***If group decisions-making generated adaptive group response exceeding the cognitive accuracy and speed of individual decisions and had a fitness consequence, then human brain size may have decreased as a consequence of metabolic cost saving.***"

"***It remains possible, then, that the high energetic cost of a heightened immune response (as a result of the deterioration of individual health) (Wells and Stock, 2020) might have been a factor in Holocene brain reduction.***"

However, Artificial Intelligence will supplement the collective intelligence of society by adding plenty of human-like elements, and it may lead to new reductions in human brain size.

* * *

Would evolution stop with Homo sapiens?

There is known that our civilization's rise has ended by many means natural selection.

As Nicholas R. Longrich (a lecturer at University of Bath, UK) has put in his article *Future evolution: from looks to brains and personality, how will humans change in the next 10,000 years?* (published on March 1, 2022, in The Conversation):

"*Unnatural environment we've created-culture, technology, cities-produces new selective pressures very unlike those we faced in the ice age. We're poorly adapted to this modern world: it follows that we'll have to adapt. Unnatural environments create unnatural selection.*"

"*In Europe, brain size peaked 10,000-20,000 years ago, just before we invented farming. Then, the brains got smaller. Modern humans have brains smaller than our ancient predecessors, or even medieval people.*

Maybe living in a large society of specialists demands less brainpower than living in a tribe of generalists. Modern humans perform fewer, more specialized roles as part of vast social networks, exploiting division of labor."

"*Domestic animals also evolved smaller brains. Sheep lost 24% of their brain mass after domestication; for cows, it's 26%; dogs, 30%.*"

"*Not everyone is psychologically well-adapted to this existence. Our instincts, desires and fears are largely those of stone-age ancestors. Modern society meets our material needs well but is less able to meet the psychological needs of our primitive caveman brains.*"

"***Could new human species evolve?*** *For that to happen, we'd need isolated populations subject to distinct selective pressures.* ***If people were culturally segregated, distinct populations, even species, might evolve***. *Still, to the extent culture divides us, it could drive evolution in different ways, in different people. If cultures become more diverse, this could maintain and increase human genetic diversity*".

"*Our genes are being curated by computer, just like our playlists. It is hard to know where this leads, but I wonder if it's entirely wise to turn over the future of our species to iPhones, the internet, and the companies behind them.*"

"***As technology and culture enter a period of accelerating change, our genes will too****. The most interesting parts of evolution are about what's happening right now, our present-and our future.*"

* * *

Digital revolution and its impact on cerebral development

There are 10,000 genes (which are 1/3 of the total number of genes) that are connected to the human brain and nervous system functioning.

Martin Korte, PhD published in Dialogues in Clinical Neuroscience, Vol 22, No2, 2020 his study "*The impact of the digital revolution on human brain and behavior: where do we stand?*"

This study "*outlines the current results of neuroscience research on the possible effects of digital media use on human brain, cognition, and behavior.*"

The author explains:

"*In the UK alone, according to data collected by a regulatory agency for communication (Ofcom), 95% of people aged 16 to 24 years old own a smartphone and check it every 12 minutes. Estimates suggest that 20% of all adults are online more than 40 hours per week.*

Nearly 4.57 billion people worldwide have access to the Internet, according to data published December 31, 2019."

"*The results published implicate intensive digital media use in reducing working memory capacity; psychological problems refer to aspects from depression to anxiety and sleep disorders; there is an influencing of the level of the text comprehension while reading on screen. Digital environments lead to superficial engagement in text analysis and reduce attention span and generate attention-deficit hyperactivity disorder (ADHD) is much higher than it was 10 years ago.*

***The use of fingertips on touchscreens changes cortical activity in the motor or the somatosensory cortex**. It was already known that cortical space assigned to the tactile receptors on fingertips is influenced by how often the hand is used. For example, string instrument players have more cortical neurons of the somatosensory cortex allotted to the fingers they use in playing the instrument.*"

"***Intensive touchscreen use can reorganize the somatosensory cortex and it occurs at the expense of other motor coordination skills**. Other aspects are effects on language, cognition, and perception of visual objects in the developing brain.*

Gomez et al., showed that details of the development of the visual system can be affected by the content of digital media. The data indicate that digital media use can lead to a unique functional and long-lasting representation of digital figures and objects even decades later."

"*The data shows the tremendous plasticity of the brain to add new representations for novel classes of objects to the higher visual area or whether object representation from **intensive digital media use might have negative consequences of competition for cortical space**.*

*A correlation between time spent with digital media and lower cognitive empathy with other humans has been reported. Also, early extensive screen use in preschoolers can have dramatic influence on language network. Overall, **a clear correlation was observed between intensive early childhood digital media use and poorer microstructural integrity in white-matter tracts, especially between the Broca and Wernicke areas in the brain**. Lower executive functions and lower literacy abilities were observed. Reading skill might be compromised if fiber tracts between the language areas are not developed to their full extent. Functional connectivity was decreased within the language network and increased between visual, DMN, and cerebellar networks, **suggesting decreased strain on the language network afforded by pictures and visual imagery**.*

Intensive media usage can stimulate the ***growth of new synaptic connections, while at the same time eliminating neuronal synaptic connections that are used less often****. The storage of relevant facts may be compromised, if the input to a neuronal network in a particular brain area exceeds its limit of processing power.*"

"*The results suggest that subcontracting some simple mental searches to internet cloud storage and relying on search engines instead of memory systems in our own brain* ***reduces our ability to memorize and recall facts in a reliable manner****.*"

The author concludes:

"*Will these new technologies build the perfect bridge to even more sophisticated forms of cognition and imagination, enabling us to explore new knowledge frontiers that we cannot at the moment even imagine? Will we develop completely different brain circuits arrangements?*"

J. Michael Cavanaugh, Catherine G. Giapponi and Timothy D. Golden published a study "*Digital Technology and Student Cognitive Development: The Neuroscience of the University Classroom*" that appeared in Journal of Management Education on October 30, 2015.

The authors explain:

"*Digital technology has proven a beguiling, some even venture addictive, presence in the lives of our 21st century (millennial) students. And while screen technology may offer select cognitive benefits, there is mounting evidence in the cognitive neuroscience literature* ***that digital technology is restructuring the way our students read and think****,* ***and not necessary for the better****.* ***Certain cognitive skills are gained while other "deep thinking" capabilities atrophy*** *because of alterations in the neural circuitry of millennial brains.*"

Professor Werner Leodolter of Applied Management in Healthcare, Graz, Austria gave an interview about his article *Digital transformation and the subconscious* that appeared in Management Matters, Volume 17, Issue 5, 2017.

He explains:

"***Mental and algorithmic models are parts of the subconscious mind*** *of your organization and significantly influence and drive the decision taken. Technology changes the cognitive process of individuals as well as organizations. Cognitive computing already directly addresses cognitive processes which were a privilege of humans until now.*

Perception is changed by virtual reality, augmented reality, mixed reality, social media, ubiquitous sensors, and drones. ***Analysis and interpretation are changed by pattern recognition, realtime analysis, and several forms of AI****.*

We simply do not know where digital transformation will lead us. Uncertainty is not vanishing from earth because we have big data and prediction algorithms. So, we need new models of thinking and scrutinizing decision proposals. We must organize ourselves anew to be able to cope with the challenges posed by the exponential development of AI, etc. The basic principle should be: People do not serve structures, but organizations, technology and structures serve people."

He refers to some mental operations, which after being repeated several times, become done in an "automatic manner". This **automated-type of task doing is subconscious**.

Hence, the **technology will replace much of our "automatic doing" (repetitive doing) or the subconscious part of it**; would it eventually affect, too, to the unconscious thought?

* * *

Another view on swarming

The self-organizing elements could mount a "natural self-complexity'. Swarming is such a type of complexity that implies an emergent behavior that remains in the range of "collective intelligence', while it remains strictly connected to a collectivity or a group. It solves the problems of the group as a whole, while the individual does not develop any "intelligent behavior". For it, there occurs a **recursive behavior of the**

group, where it is submerged. It has been estimated that human swarming was present in our species until 30,000 years ago or less.

However, **swarming is now an increasing part of computational development, and it will cause a certain interference with human subjects and their collective intelligence**.

AI is an exogeneous capacity characterized by a high speed of computing. When added to our brain, it will remove many tasks and associated skills, which our brain uses to perform. This will influence the individual intelligence in the same way every swarming does.

The replaced skills would turn atrophied, influencing the general functioning of the neuronal networks and their interconnectedness. **Such a powerful exogeneous capacity will affect our brain architecture anew, like during pre-history, when other strong forcing factors from the environment significantly did**.

However, Dr. Louis Rosenberg in his paper *Human swarms, a real-time method for collective intelligence*, published in July 2015 (DOI:10.7551/978-0-262-33027-5-ch117), provide some valuable ideas about the role that human swarming assisted by computational systems could offer to very complex problems of our society.

"*Swarms are unifying, enabling groups to find commonality among them and converge on solutions that optimally support . It is also a return to our roots. UNU is the computational glue that allows people to work together in swarms that can answer questions, generate ideas, even express opinions. In such a way, real-time synchronous control is enabled across a swarm of distributed networked users. In all cases, the predictions made by swarms were more accurate than the predictions made by the individuals who comprised the swarms. In fact, the smarms consistently performed better than even the most skilled individuals in each group. And because we can't prevent researchers from creating pure AI technologies that may rival humanity, it's our hope that human swarming will offer a way to stay one step ahead of the machines*."

During several tests, "*the average individual achieved 53.7% correct, while the average swarm achieved 76.7% correct, corresponding to an estimated IQ increase of 14 points*."

Let's scrutinize the physics that may be behind the possible mechanism of the swarm phenomenon.

Alberto Ambrosetti, Paolo Umari, Pier Luigi Silvestrelli, Joshua Elliott and Alexandre Tkatchenko published in Nature Communications 13, Article number 813 (2022) on February 10, 2022, the study *Optical van-der-Waals forces in molecules: from electronic Bethe-Salpeter calculations to many-body dispersion model.*

The authors explain:

"*Molecular forces induced by optical excitation are connected to a wide range of phenomena, from chemical bond dissociation to intricate biological processes that underpin vision.* ***We find that both attractive and repulsive optical van der Waals (vdW) forces can be induced by light****. These optical-vdW interactions dominate over vdW dispersion in the long-distance regime,* ***showing complexity that grows with system size****. It suggests the ability to selectively activate molecular vibrations by photoabsorbtion.*"

"*In conclusion, we induced an effective model for semi-quantitative analysis of vdW forces in optically excited nanostructures and extended systems.* ***Optical excitation*** *(*as well as its photoacoustic derivative, my note*)* ***can be exploited to induce both attractive and repulsive long-range forces between well-separated molecules****.* ***The complexity of these forces grows rapidly with system size****, due to intricate geometrical features of collective atomic forces.*

Depending on the specific application, optical vdW forces can be potentially turned from attractive to repulsive, further altering both their strength and power-law decay."

Matan Mussel and Matthias F. Schneider in their study *Similarities between action potentials and acoustic pulses in a van der Waals fluid*, published online 21 February 2019, and in Scientific Reports, explain:

"While photoinduced response is predominantly identified with chemical bond distortions, recent experiments provided direct evidence of unconventional van der Waals-like interactions between Rydberg atoms at a Forster resonance. ***It suggests experimental ability to induce and control molecular forces by optical excitation, targeting biological systems****."*

There is known that ***steady emission and absorption of biophotons observed in all biological systems suggest that biophotonic-emission (light) may provide a supplementary mechanism for information exchange, in addition to electro-chemical signals***.

"Acoustic pulses (as a derivative of the photoacoustic effect) in lipids can be excited by various types of stimulation like mechanical, electrical, chemical, optical. Non-linear acoustic pulses arise naturally from a minimum set of well accepted physical assumptions: mass, momentum, and energy conservation laws near a phase transition."

In the medical field the **photoacoustic effect** is exploited as an **ultrasound-mediated biophoton activation** (produced internally and externally) and processing in imaging.

Bruno Bordoni, PhD DO, Fabiola Marelli, PhD Do, Bruno Morabito, PhD DO, and Beatrice Sacconi, MD published (in Journal of Evidence-Based Integrative Medicine, online 2018, feb.1) a study *Emission of Biophotons and Adjustable Sounds by the Fascial System: Review and Reflections for Manual Therapy.*

The authors explain: "*This article reviews the literature on the emission of biophotons and adjustable sounds by the fascial system.*"

"*The fascial system has various tools for communicating inside and outside the body.* ***These communications allow rapid adaptation of bodily functions, from posture to metabolic functions, from vocal expression to psychological and pain perception****. The fascia communicates with liquid tools, through ducts known as the Bongham ducts, a network that carries information to all parts of the body, independent of lymphatic and blood systems.*"

"*The Ultraweak Photon Emission (UPE) occurs without the presence of external stimuli, while the induced one is caused by biotic factors, and abiotic factors like the temperature and gas present in the environment, mechanical and light stimuli, ionizing radiation.*"

"***The electric transition of electrically excited species during the oxidative metabolic processes is accompanied by the emission of photons**. The presence of free radicals or reactive oxygen species (ROS), corresponds to the presence of UPE. **Electrical activity related to excitation of the membrane of the cellular tissues is probably more important in triggering emission***."

In another study (*X-ray induced bio-acoustic emission from cultured cells*, published by BioRxiv on June 16, 2021), the authors Bruno F. E. Matarase, Hassan Rahmoune, Nguyen T. K. Vo, Colin B. Seymour, Paul N. Schofield, and Carmel Mothersill explain:

"*We report the first recorded **acoustic signal captured from a collective pressure wave response to ionizing radiation**. We show that at doses of X-ray irradiation capable of producing the release of bystander effect-inducing activity, both **cell types emit a characteristic acoustic signal for the duration of the irradiation pulse***." "*A cumulative body of evidence suggests that in addition to biochemical factors mediating the radiation-induced bystander effect (RIBE), biophysical elements can contribute to RIBE. There is an intrinsic relationship between electromagnetic forces (mediated by photons) and acoustic vibration (mediated by phonons). On exposure to ionizing photons, inner-shell electrons are excited, generating photoelectrons, Auger electrons, or electromagnetic radiation. **These processes lead to a transient thermoelastic expansion of the biological structure generating the pressure waves of an acoustic emission** (Garcia et al., 1988)*."

"*We demonstrate here that **irradiation with 1Gy of 6MeV X ray photons can cause cultured cells to emit characteristic sound waves, and this acoustic signal might trigger the bystander response in the neighboring cells***."

Another study (*Quantifying Biophoton Emission from Human Cells Directly Exposed to Low-Dose Gamma Radiation*, published by Dose-Response An International Journal. April-June 2020 1-7) by Jason Cohen, Nguyen T. K. Vo, David Robert Chettle, and Fiona E. McNell shows:

"*For the first time, gamma radiation was shown to induce biophoton emission from biological cells. Our group has further shown that photons emitted from beta-irradiated cells can stimulate physical signals inducing RIBEs in bystander cells.*"

As exogenous factors, the gamma and X rays of cosmogenic origin generate in the terrestrial atmosphere (but X-rays are fully absorbed by the atmosphere) showers of secondary particles, like the C14 isotopes, which are low-dose ionizing radiation.

Hence, the **Earth's biota may be influenced by C14 isotopes**, causing biophoton emission and acoustic phonons. Pulses of C14 isotopes may trigger in biota the stimulation and activation of the van der Waals forces, temporarily influencing the swarm phenomenon in various organisms.

It implies that the swarm phenomenon is mostly generated by endogenous factors, where some environmental variations (like temperature, gaseous pressure, and background radiation) may also temporarily interfere.

In sum, **the swarm phenomenon has at its fundament the van der Waals forces, which becomes stimulated** (augmented) **by signals** (biophotons and other photoacoustic resonant excitation) **emitted by the biological and non-biological systems**. Such signals cause the rising of attractive and repelling forces, which are distinctly used by each single swarm mechanism.

As the experiments proved, the **van der Waals attractive/rappelling force manifests only when a particular resonant excitation occurs within at least two systems, and it augments proportionally with the number of systems becoming clustered by their individual excitation**.

When the excitation stops, the van der Waals manifestation ceases, too, and the swarm phenomenon perishes unnoticed. The same phenomenon similarly arises for inorganic and organic matter.

Recent research at the University of Helsinki (Finland) (August 30, 2022) shown that people's brains become activated in a similar and simultaneous way during social interaction. Such inter-brain neural synchronization has been linked with empathy and cooperation in face-to-face situations. The mentioned research now shown that **the inter-brain phase synchronization can occur without the presence of the other person but in online interaction, like during cooperative online gaming**.

Here, the **synchronization may be the result triggered by** the same van der Waals force activation because the **van der Waals field is a continuum**, and the distance may turn meaningless. This is so, because **the subtle force manifested between particles (or organisms) is mediated by quantum fluctuation of charge**. It is a quantum phenomenon manifested at nanoscale level.

How such a no-physical--barriers-synchronization affect our social mechanisms?

It shown that phase synchronization of oscillatory activity occurs online real-time joint coordination without any physical co-presence or video or audio connection.

Hence, **synchronization** (as a quantum manifestation at nanoscale) **between isolated human brains can occur online in the same way it occurs during face-to-face interaction**.

* * *

Evolutionary changes of the oculomotor system and their implications for analyzing the reality

In my opinion, the fundamental aspect about our future refers to the **speed of processing of the details embedded in the environment**. With the advent of language, 30,000 to 23,000 years ago, the human brain added new means to decompress reality, leading to unveiling more and more of its details. Simultaneously, our capacity to combine and/or compose the unveiled details into more complex results also gradually increased, following the cerebral development of the Frontoparietal Synthesis.

Bela Weiss, Balazs Knakker, and Zoltan Vidnyanszky published in Scientific Reports on 2016 May 27 a study *Visual processing during natural reading.*

The authors explain:

"*Increased letter spacing* (1.5 times more than normal spacing) *leads to reduced reading speed. It might be a close association between sensitivity to configural properties of orthographic information and reading skills*."

"*Neural coding of letters and words is subserved by ventral occipito-temporal visual cortical regions*." (It was also found an important contribution from centroparietal region that was developed in the last 70,000 years.)

"*We estimated how changing letter spacing alters eye movements. Visual processing load is decreased because of increased inter-letter spacing. On the other hand, saccade amplitude* (as the distance traveled by the eye between two fixation points) *were increased with larger letter spacing. Reading was fastest at the smallest inter-letter distance and decreased with the increase of letter spacing*."

"*We found that orthographic processing is reflected in FOREA in three consecutive time windows (120-175 ms, 230-265 ms, and 345-380 ms after fixation onset*)."

From the experimental information of this study, I concluded:

-saccade peak velocity declines as latency increases.

-visual processing load is decreased because of increased inter-letter spacing.

-saccade amplitude increases with larger letter spacing, showing a decreased latency.

Hence, **smaller letter spacing (more visual input details) increases the latency**. Or one can say **the latency increases unveils a reality with more details**.

However, the space of the latency is filled with information produced by the unconscious that was previously unseen by the consciousness.

Thus, we deal here with the "details" provided by the unconscious, and which become conscientized in the space of the generated latency.

Are they the unseen details of reality? Such details are provided by so called nondeclarative memory that is expressed through performance instead of recollection. It gradually extracts the common elements inhabiting a series of separate events. This undeclared memory is very large because the unconscious is immensely larger than the conscious. Hence, it has myriad unconscious ways to respond to visual tasks.

However, such unconscious information refers to past events and is characterized by a type of short-memory that fills the space of latency. While such information is not available to consciousness, it still was recorded by the unconscious from the reality and becomes inserted into the space of the latency. Thus, it is a novel information for consciousness that was not available before for conscious processing.

A study (*Unconscious associative memory affects visual processing before 100 ms*) published in February 2008 by Maximilian Chaumont, Valerie Drouet, and Catherine Tallon-Baudry indicates:

"***No conscious knowledge regarding the identity or the spatial properties of the search arrays could ever be shown*** *(Chunk & Jiang, 1998, 2003). In sum, these results show that human observers quickly develop an unconscious memory of the repeated displays and use it automatically to guide attention when viewing previously encountered visual scenes*."

The mentioned study contradicts other scientific information (studies on the reading mechanism of texts) that indicates that the unconscious introduces novel information about the identity and spatial properties of the elements displayed within the space of latency.

The average visual perception of a scene takes from 100 to 200 ms. Here, speaking and listening cause an additional latency that amounts on average to 150 ms. Thus, the processed latency extends from 200 ms to an average of 350 ms (I cited a study by Gerry T. Altman published in June 2011 in ACTA Psychologica -Elsevier).

Stephan Levinson of Max Plank Institute for Psycholinguistics in a study published in 2015 explains:

"*It takes at least 600 ms for us to retrieve a single word from memory and get ready actually to say it. For a short clause, that processing time rises to 1,500 ms*."

A speaker-listener will process in, at least, double the time compared with a regular visual but silent observer.

The study on monkeys found that their latency is around 40 ms. Hence, humans, especially in the last 70,000 years, have evolved to an average latency of 200 ms, which, depending on the situation, can be increased to almost 1,500 ms.

Compared to latency found in monkeys, our latency has been increased to almost 37 times, leading to a corresponding increase in the details being observed. In all this, in the last 23,000 years, the language has played a very important role.

As one has seen above, the **increased latency produces an increased amount of details of the environment**. It is like decompression. Such unveiled details may have represented the discoveries we have made in the last 30,000 years.

However, our capacity to understand the general context diminishes as the latency increases.

Our brain organization seems to be made of contextual modules in which we compare the characteristics of various contexts. Here, every passed context influences the present and future one. It is like a fractal made of self-similar contexts.

Thus, **the means of the context are fundamental for the brain output**.

In my opinion, we have today an enormous quantity of details or discoveries, but **our understanding of their natural context remains very poor**. We are submerged into a sea of details where we cannot dive but drown.

The poor large-context capacity affects the way we understand the fundamental laws and mechanisms which regulate the Universe. In the meantime, it affects our ability to understand natural processes evolving on our planet, like global warming, pollution, ecosystems, and the expansion of our civilization.

We have problems understanding our own origin, like, until now, we have failed to understand what generated our modern brain and how fundamentally different we are from Homo sapiens. We do not really know who we are, and what kind of reasoning must be applied to ourselves?

However, in recent decades a significant increase in visualization occurred: pictures, movies, TV, videos, computer games. What will be the impact of this reversing path toward visual vs. spoken and written language?

Will cause it a reduction in the processing of the overall speed, or something else will occur?

Do we need all such details or discoveries? Would a reverse path lead to less discoveries, while improving our understanding of the surrounding context?

Another fundamental problem we face is that of the **complexities we have generated around us**. Such complexities evolve exponentially in rhythm with the path of discoveries we make and combine.

As the Science of Complexity would advise, to solve any complex problem, one must generate a (mental) complexity at least the size of the complexity to be resolved.

Would a **regeneration of human swarming**, stimulated by an ever-increasing computational capacity, be able to resolve the complexities we have generated? Possibly, such **a novel human swarming may solve at least some of the mentioned problems**.

* * *

Another digression on swarming

I like to make a digression here, where I like to present four essential elements, which may have influenced the swarming behavior in humans.

There is known that monkeys sometimes travel in large colonies of hundreds of individuals, which make a type of swarm that works well together and in tandem with other swarms.

In apes, the swarming behavior shows a significant diminishing. It persists in baboons, but it is quite weak in chimpanzees, our distant cousins.

Olaf Witkowski and Takashi Ikegami in their *study Emergence of Swarming Behavior: Foraging Agents Collective Motion Based on Signaling*, published in PLOS One, April 27, 2016, show the results of their experiments.

"*The results demonstrate that agents progressively evolve the ability to use the information exchanged between each other via signaling to establish temporary leader-follower relations. These relation allow agents to form swarming patterns that improves the agents' ability to forage for resources*."

These researchers investigated on a computerized system "*the minimal conditions for the evolution of a swarming behavior.*"

"*The individuals turn towards each other based on the signals they emit. The presence of the resource is needed to make the clustered swarming behavior appear. **If at any point the signaling is switched off, the agents immediately break the swarming formation**. A swarming behavior is only observed once the communication is turned back on. Silenced agents end up with all individuals wandering around erratically.*"

Here, I must mention that the mentioned signaling refers to a variety of means, ranging from chemical, to odor, to audio stimulation.

Another study made by Tingting Xue, Xu Li, Peter Grassberger, and Li Chen with the title *Swarming transitions in hierarchical societies*, published in arXiv: 1912.01313v3 (physics.bio-ph) on 26 Oct 2020 explain: "*A major concern here is the nature of swarming transitions between the ordered and disordered states of movement*."

They used the Viscek model inspired by statistical physics.

"*In this work, we fill the gap by introducing a hierarchical swarm model, Viscek model, and investigate the impact of the hierarchy* ***on nature of order-disorder transitions****. By tuning the swarm from egalitarian (organization) to despotism, we show numerically that the swarming transition changes non-monotonically from a weak first order PT to stronger one for intermediate levels of hierarchical impact, and finally the discontinuity shrinks and disappears*."

In short, this study demonstrates that the swarming turns from weak (for an egalitarian swarm society) to strong, when the hierarchical organization develops. Here, the introduction of **hierarchically displayed leaders improves the performance of the swarm**. **When the swarm's hierarchical organization climbs to leaders' despotism** (absolutism or no communication with the rest of the members)**, the swarm behavior vanishes**.

However, this experiment at the Xian University (China) was of a mathematical nature, thus, the meaning of absolutism refers to total lack of communication between the leader and the rest of the members of the swarm. Here, the swarm vanishes.

This result is similar to that of the previously quoted study, where the **silenced signaling causes the dissolution of the swarm**.

Now, I will discuss the third essential element in the mechanism of the swarm. It refers to the human language advent. When around 30,000-23,000 years ago humans started to talk a coherent language with verbs and adverbs, this new type of communication affected the previous type of communication that made use of all sense's inputs. It altered or replaced much of the previous way humans have communicated and eventually it discontinued the fulness of that communication, disbanding the swarm organization.

The fourth element that may have disrupted the human swarm behavior was the change in the oculomotor system that significantly increased the saccade latency, making humans see more details of the environment. **Such addition of information may have disturbed the traditional way the swarm mechanism worked, introducing new means of interaction**, which in part replaced the swarm function.

Could swarming be re-created in humans? This question implies a lot more studies, which must define which exact signals in various animals (biota) activate the van der Waals forces. It must find how such signals can be stimulated in humans.

Even then, careful care must be employed in stimulating swarming in humans, because some detrimental aspects may arise.

* * *

Some ideas on our future

Now I would like to quote the collapse expert Luke Kemp, who wrote extensively on this subject. I am referring to his paper *Are we on the road to civilization collapse?* published on 18 February 2019.

He says: "***civilizations are often responsible for their own decline and their self-destruction is usually assisted***."

"*Collapse can be defined as a rapid and enduring loss of population, identity, and socio-economic complexity. So, collapse may be a normal phenomenon for civilizations, regardless of their size and stage.*"

The collapse may be triggered by climate causes and excessive environmental degradation.

Wealth and political inequality can be central drivers of social disintegration. Inequality undermines collective solidarity that is manifested by swarm intelligence of the society.

"*Societies eventually collapse under the weight of their own accumulated complexity. Societies are problem-solving collectives that grow in complexity in order to overcome new issues. However,* ***the returns from complexity eventually reach a point of diminished return. After this point, collapse eventually ensue***."

Wealth and health inequalities would become worst as the degradation evolves.

Another aspect is **an "energy cliff", where the degradation goes to a point where current societal levels of affluence can no longer be maintained**.

"*People are increasingly specialized and disconnected from the production of food and basic goods. And a changing climate may irreparably damage our ability to return to simple farming practice.* ***The higher we climb, the larger the fall. Any collapse risks to be permanent***."

"***Assistance in our self-imposed ruin will not come from hostile neighbors, but from our own technological powers***."

However, another book by Craig Collins "*Four Reasons Civilization Won't Decline: It Will Collapse*" (published 13 March 2020 by Counter Punch) presents a different eventuality for our immediate future.

In brief, it indicates that "*It's incorrect to extrapolate that the fact that we've made progress is a prediction that we're guaranteed to make progress*" (cited from Steven Pinker work by the author).

"*The progress of the past was built by sacrificing the future-and the future is upon us*."

He cites John Michael Greer's book *The Long Descend* by saying "*The same pattern repeats over and over again in history.* ***Gradual disintegration, not catastrophic collapse is the way civilization ends***. *It takes an average of 250 years for civilization to decline and fall*."

"*Today, the rich easily exploited oilfields and coalmines of the past are mostly depleted. And, while there are energy alternatives, there are no realistic replacements that can deliver the abundant net energy fossil fuels once provided.*"

"*The economy of industrial society is capitalist. Production for profit is its prime directive and driving force. The unprecedented surplus energy supplied by fossil fuels has generated exceptional growth and enormous profits over the past two centuries. But in the coming decades, these historic windfalls of abundant energy, constant growth, and rising profits will vanish.*"

"*As profitable sources of production dry up, capitalism will be compelled to turn a profit by consuming the social assets it once created.* ***By cannibalizing itself, the profit motive will exacerbate industrial society's dramatic decline****. Catabolic capitalism will profit from scarcity, crisis, disaster, and conflict. Capital will flow toward lucrative ventures like cybercrime, predatory lending, and financial fraud; bribery, corruption, and racketeering; weapons, drugs, and human trafficking.*"

"*Once disintegration and destruction become the primary source of profit, catabolic capitalism will rampage down the road of ruin, gorging itself on one self-inflicted disaster after another.*"

We should ponder about the reality we have built around us, where "*in terms of biomass, 96 percent of all the mammals on Earth are livestock; only 4 percent are wild mammals. Seventy percent of all birds are domesticated poultry, only 30 percent are wild. About half the Earth's wild animals are thought to have been lost in just the last 50 years.*"

When our society will disintegrate, as expected, how this mentioned animal balance be affected? Would there be an existential treat related to food security?

How the aging of the population in the developed countries would influence the value of its collective thinking and future global problem-solving?

For the last word, I like to quote David H. Wolpert article *A sliver of reality* published by Aeon.co in September 2022.

The author in his 10th question says "*Is there any way that we could imagine testing whether our future science and mathematics can fully capture physical reality? From a certain perspective, this question might appear to be a scientific version of a conspiracy theory, writ large.*"

"*More speculatively, as our cognitive abilities grow, we might be able to establish the existence of what we can ever conceive of through observation, simulation, theory, or some other process. In other words,* ***it may be that the feedback loop*** *between our extended minds and our technology* ***does let us break free of the evolutionary accident that formed our hominin ancestors' brain***."

"*Cognitively superior children of ours will be here within the next century. Presumably,* ***we will go extinct soon after their arrival*** *(like all good parents making way for their children). So, as one of our way out the door, as we gaze up at our successors in open-mouthed wonder, we can simply ask our questions to them.*"

At the end of this section of the book I would like to reassert **the mechanism which propels the Science of Complexity because our civilization evolves as a growing complexity**. The way we would understand this mechanism, its wisdom, would be the way we would surf and resolute into our future.

I would quote from a paper uploaded online by Steve Kaisler and Greg Madey in 2008 for the University of Notre Dame(HICSS-42, 2008) because it is significant when one talks about the future.

As Pogo said: "*We have met the future and the future is us, because humans continuously evolve, not physically, but emotionally, cognitively, etc.*"

Our future depends not on us, but on very many other systems, where we are a smaller part of them. We are embedded in an infinite chain of systems.

"*Within a complexity, we are the low-level entities, which constantly interact.* ***The final product that results is a nonlinear interaction rather than planning or design, while such a result cannot be known a priori***."

At every level of complexity entirely new properties would appear.

"***Emergent refers to the state of being in continual process, never arriving, but always in transit. There are no points of theoretical stasis, only emergent regularities, and those regularities are always shifting and evolving***."

"*Changes in behavior can occur due to changes in the environment without corresponding changes in the agent. Changes in the agent can affect changes in the environment.*"

"*Signs of "purpose", and even systems that would exhibit "purposive" behavior, can come into existence simply because indeterminate processes, i.e., need not involve minds at all.* ***A "purposive" behavior does noy depend on any representation of "purpose" within the agent***."
"*Ambiguity and paradox abound in complex adaptive systems which use contradictions to create new possibilities to co-evolve with their environment.*"

"*The relationships between the agents are generally more important than the agents themselves. Agent changes often occurs in response to feedback from their actions.*"

"*There is no hierarchy of command and control in a complex system, there is no planning or managing, but there is constant re-organization to find the best fit with the environment. The system is continuously self-organizing through the process of emergence and feedback.*"

"***We continuously adapt to the changes around us, but we do not learn from the process. Only the Complex Evolving System (CES), or the system itself, can learn from each change, enabling them to influence their environment, better predict likely changes in the future, and prepare for them accordingly****. Such a system-level behavior is what intrigues us.*"

The origin of human quantification and its relation to human language

The aim of this paper is to define when linearization and quantification begins to manifest in the modern human brain.

It also opens a discussion on the reverse effect of quantification that can lead to novel qualitative gains.

John E. Opfer, and Robert Siegler authored (2012-02-29) a study *Development of Quantitative Thinking* (https://kilthub.cmu.edu/articles/journal_contribution/Development_of_Quantitative_Thinking).

The authors explain: "*For understanding development of quantitative thinking, the distinction between nonsymbolic and symbolic thinking is fundamental. Nonsymbolic quantitative thinking is present in early infancy, culturally universal, and similarities include the ability to represent and compare numerosities, the representations being noisy and increasing logarithmically with actual quantity and neural correlates of number representation being distributed in homologous regions of frontoparietal cortex. Symbolic quantitative thinking, in contrast, emerged recently in human history, differs dramatically across cultural groups, and develops over many years. As young children gain that range, they initially map them to logarithmically-compressed mental number line and later to a linear form. This logarithmic-to-linear shift expands children's quantitative skills profoundly, including continuous and discrete quantities, categorizing numbers by size, remembering numbers, and to estimate and learn answers to arithmetic problems. Thus, while nonsymbolic quantitative thinking is important and foundational for symbolic numerical capabilities, the capacity to represent symbolic quantities offers crucial cognitive advantages.*"

David Landy, Colin Allen, and Carlos Zednik published the study *A Perceptual account of Symbolic Reasoning* in Frontiers in Psychology, Volume 5, 201, https://doi.org/10.3389/fpsyg.2014.00275).

The authors indicate: "*How do people reason arithmetically, algebraically, and logically? One well-known answer to this question holds that the human mind trades in inner symbols that amodally represent abstract arithmetic, algebraic, and logical propositions, and manipulates these symbols according to internally represented mathematical and logical rules. On this traditional view the inner takes precedence over the outer: notations on paper, computer screens, and classroom blackboards are involved in mathematical problem-solving only insofar as they are translated into corresponding mental structures and processes*."

"*Whatever or not the given symbols described a valid equation, the expressions contained both additions, which stipulates that multiplications precede additions. Related research has indicated that spatial layout impacts application of the order operations rules when calculating, when creating story problems, and when working in programming languages*."

"*The ability to follow operator-precedence rules is just one manifestation of the capacity for symbolic reasoning: the capacity to manipulate arbitrary symbolic tokens according to abstract mathematical and logical rules. Rather than represented rules, symbolic reasoners make their mathematical judgement using perceptual processes that have no obvious link to the following of formal mathematical rules. Instead, symmetry in, and otherwise perceptually organize symbolic notations as that are experienced in the environment. In contrast, when notations do not afford physical manipulations or perceptual processing, symbolic reasoning may involve processes of visual, aural, and event tactile imagination. Although symbolic reasoning can be therefore become internalized, it remains rooted in mechanisms close to sensorimotor periphery*."

"*Analogous to the syntactic approach, computationalism holds that the capacity for symbolic reasoning is carried out by mental processes of syntactic rule-based symbol-reasoning system that is functionally encapsulated from dedicated and modality-specific-sensomotor modules*. "Other theories support that "sensorimotor processing can be safely "abstracted away." "*On all computationalist accounts, when individual is confronted with a symbolic reasoning task such as a natural-language word-problem or a format reasoning problem expressed in the notational formalism of algebra, calculus, and logic, the perception of notations in the environment causes a tokening of equivalent symbols and expressions of "Mentalese" (Fodor, 1975). These mental symbols and expressions are then operated on by syntactic-rules that instantiate mathematical and logical principles*."

"*Computationalist, and semantic processing accounts share the assumption that processes of perception and action play a relatively limited role in the process of symbolic reasoning. They suppose that processes of perception and do little other than mediate between notational structures in the external environment, and the internal structures and processes in which symbolic reasoning* ***really*** *occurs*."

"*Like many other kinds of problem solving, the process of symbolic reasoning can be seen as a chain of transformations that links input and output representations, each of which changes its format and/or semantic structure. The resulting representation is a simplification of the original problem*."

"*The principal role of sensomotor processes- the processes that govern the perception of and physical interaction with public symbols and expressions-is simply to provide inputs and carry outputs from those internal structures and processes that are ultimately responsible for performing all substantial steps in mathematical or logical problem-solving chain. On a constitutive account, sensorimotor mechanisms not only translate the problem, but they are also involved in the transformation that substantively solve it*."

Edna Orr and Ronny Geya authored the study *Infant Behavior and Development* (published in Science Direct, Volume 38, February 2015, pages 147-161).

The authors explain: "*Symbolic play and language are known to be highly interrelated, but the development process involved in this relationship is not clear. Three hypothetical paths were postulated to explore how play and language drive each other.*"

"*Symbolic play, or pretended play, and language are known to be highly interrelated. Both relay on representational capacity, namely, employing one element as a signifier to represent another element. In pretended play, objects or situations are used or performed in a way that does not exist in the immediate reality., whereas in language, a vocal symbol (a word) represents an internal meaning that is related to entities or events in the real world. Furthermore, both behaviors, pretend play, and language, share a similar developmental architecture, progressing from the most basic to more advanced forms. The transition from basic to advanced forms is coherent symbolic act.*"

"*Symbolic play begins at the pre-symbolic play level when infants are capable of recognizing the real relationship between familiar objects and their related actions. Generally, the transition to symbolic play is evident when the infant uses sound effects or gestures, indicative of the referent behavior. As symbolic ability increases, infants become more capable of combining mental representations of several signifier-signified relationship into sequences or organizing them into a hierarchical order. Infants reach the pre-symbolic level between 8 and 11 months of age, and the first milestone of symbolic play is typically evident at around 11-12 months of age.*"

"*Language, like symbolic play, begins with basic forms. The development of language begins with babbling (vocalizations consisting of syllable repetition). Babbling emerges before the ability to talk, generally before 10 months of age. Babbling is also shown to follow rhythmic motor activity, such as repetitive arm movements that accompany repetitive vocalization. According to this view, babbling may represent a vocal play designed to train the motor challenges that are involved in speaking.*"

Pre-symbolic thinking is common in all animals. The "**abstract**" thinking first appeared in Homo sapiens brain around 70,000 years ago. It evolved gradually and became better represented in cave and rock painting sometimes between 40,000 and 30,000 years ago. The last abstract cave painting, Altamira Cave (Spain), is dated 34,000 years ago.

However, the first symbolic cave painting appeared 30,000 years ago. It switched into symbolic thinking that developed in parallel with language. Hence, **human linearization and quantification started to manifest around 30,000 years ago and corresponds with an archaeologically demonstrated "language revolution."**

It is something people are not born with. Abstract thinking develops during childhood and continues to evolve in the teenage years.

"Rostral prefrontal cortex (RPFC) has increased in size and changed in terms of its cellular organization during primate evolution. In parallel emerged the ability to detach oneself from the immediate environment to process abstract thoughts and solve problems and to understand other individuals' thoughts and intentions. Abstract thoughts can be broadly defined as thoughts that are self-generated and stimuli-independent, in contrast to stimulus-oriented, perceptually-derived, information." (Iroise Dumontheil, doi.org/10.1016/j.dcn.2014.07.009). Abstract thinking, or abstract reasoning, is the ability to understand complex concepts, while real, are not connected to a concrete experience, or objects, or people. It is a type of higher-order thinking that relates to symbolic or hypothetical aspects.

Young children start to develop abstract thinking during their "pretending play", and it begins to manifest around the age of five. But **abstract thinking becomes more apparent around the age of 12**. Abstract thinkers focus on the reason why a task should be performed.

As it seems, the introduction of verbs and adverbs occurred sometimes between 40,000 and 30,000 years ago, and this addition introduced the dynamics of combination and manipulation of objects, as an essential capability of symbolic and quantitative linear manipulation.

The quantification starts in pretended play and involves one digit or one object manipulation, evolving to two and/or more objects. At this point, the **bubbling becomes language, and the composition becomes a computation**.

Undoubtedly, **language is a result of mental quantification**.

One "interesting" mechanism "comes into play when the qualitative increase in some entity, usually population, reaching a certain threshold, gives rise to a qualitative change in the structure of a society.

In his book *Science of Logic*, Georg Friedrich Hegel said: "*It is said that there are no sudden changes in nature, and the common view has it that when we speak of a growth or a destruction, we always imagine a gradual growth or disappearance. Yet we have seen cases n which the alteration of existence involves not only a transition from one proportion to another, but also a transition, by a sudden leap, into a qualitatively different thing: an interruption of a gradual process, differing qualitatively from the preceding, the former state.*"

Hence, we have not only the transition from quality into quantity, but also, the reverse transition from quantity to quality.

Robert L. Carneiro in his study The transition from quantity to quality: *A neglected causal mechanism in accounting for social evolution*, published in Anthropology, on October 24, 2000 (97(23) 12926-12931 (doi.org?10.1073/pnas.240462397) said:

"*In my own professional field-anthropology-the transition from quantity to quality has received only limited recognition. Its application has been chiefly in attempts to explain how, during the course of hominid evolution, the human brain became able to engage in the symbolling behavior underlying the production of speech, and with that, to be able to generate culture.*"

Leslie A. White, an anthropologist, suggests: "*transition from quantity to quality may have been the most plausible explanation for the origin of the capacity for culture.*"

Misha Titlev, another anthropologists, said: "*... it may be that a primate brain, which is normally less than 900 cc., stand for a mentality that is incapable of true symbolization... (but) any normal primate brain, above 1000 cc is probably fully capable of using symbolic speech and other features of algebraic mentality.*"

General claim in modern science "*is that a process of gradual anatomical change can at certain points yield "sudden functional advantages that will lead to qualitative different patterns of behavior in a species*."

The author, Robert L. Carneiro, concluded: "*If a certain dislocation of the normal working of a society does not exceed a certain point, the ordinary operation of the existing institutions of that society eventually will restore it to its former state, with no permanent changes in its structure having occurred. But if the disturbance is of sufficient advancement, the structure itself will change, and the society will not be able to return to its previous state*."

However, the dialectic is the underlying philosophical fundament and mechanism of Taoism, where opposites and contradictory aspects are presented in all things (yin-yang), and the motions manifested in nature are the movements of something toward its opposite. Its founders have been in the fifth century BCE and are represented by Lao Tzu, Chuang Tzu, and Lieh Tzu.

The Chinese dialectics extended to opposing Confucianism (culturally and politically), Buddhism (as the Dialectical Mandala Model of Self-Cultivation that employs the Buddhist tetralemmic dialectic which goes beyond the limitations of dualistic and reductionist logic), and Zen Buddhism. It instantly extended regionally to Korea, Japan, and Vietnam philosophers and cultures. It also influenced the Hindu logics.

Fiske (2002) criticized "collectivism" as "*an abstraction that formalizes our ideological representation of the antithetical other, a cultural vision of the rest of the world characterized in terms of what we are not*." (p.84).

The current dialectic thinkers, like Joost Kircz and Marcel van der Linden, (in their study *Complicating the Quantity-Quality Transition*, published in Critique, 2021) indicate that the transition is in fact very complex. They bring various examples where the transition causation shows that ***"quality" is better defined as the "emergent" behavior of a collection of parts (elements) defined as "quantities***."

Undoubtedly, **the current path toward "quantification" will bring about significant "cultural changes in society**. On the other hand, the

quantitative changes will force society to adapt by changing, at least in part, social rules. Quantitative changes can trigger in themselves other qualitative changes without quantitative involvement. All changes progress toward some critical points: too much, or too little. Hence, the transition will be affected by which of these two limits will be reached.

I have estimated that **we are in a process of "continual linearization" that is the natural outcome of our modern brain anatomical changes**.

Symbolization became an innate to our modern brain because it currently manifests before a child brain reaches one year old.

Is this anatomical change reversible?

The other fundamental question regards the "latency" that the "linearization" develops. How much "latency could our brain admit without severely impeding neural speed of processing and ultimately our survival capability?

Kosuke Itoh, Naho Konoike, and a group of other researchers from Niigata University (Japan) published in Scientific Reports a study (*Cerebral Cortical Processing Time Is Elongated In Human Brain Evolution*) (doi.org./101038/s41598-022-05053-w) in December 2022, in which they indicate that human length of auditory pathway increased to double compared to monkeys. It means, auditory human "latency" has doubled in the recent course of human evolution.

"*An elongation of the auditory cortical processing time entails both merits and demerits. Slow processing is disadvantageous in situations where fast identifications of sound are required. On the other hand, a widened processing time has the advantage of allowing the auditory input to be integrated over a longer time window, which is beneficial for analyzing the time-varying temporal structures of sound. The evolution of speech in humans was supported by the enhanced capacity of sound perception that was made possible by the longer cortical processing time. The longer-range neural communication between the two cerebral hemispheres is constrained in speed as brain size increases, because of the limit of the action potential propagation speed. Brain enlargement would also increase the transmission delays in long-range intra-hemispheric connections between homologous cortical areas*."

A similar process occurred in all human sensorial capacities.

Here, I have to explain the mechanism that produces "latency".

A low entropy production shows a higher process of dissipation that brings more linear order to brain functioning. As for flows subject to wall neural tract friction, energy is dissipated when the surfaces of the flow and walls rub together. Friction increases the thermal energy of the circuits (tracts), and this energy becomes wasted, or it becomes ceded to the environment. The longer the neural circuits (tracts) become, the more signal friction occurs, which slows down the transmission speed of the signal.

The dissipation increases the stability of the system. Transport-dissipation reduces the transport capacity. In the meantime, dissipation tend to discretize, a process that linearizes the continuity of the signal flow.

Hence, "latency" has a complex role in the brain functioning, producing linearization and linear-order. A slow flow of neural signals, provided by perceptual senses, allows attention to analyze an increased number of details which are embedded in the flow deeply. It generates what I call the "discovery" of embedded details.

However, there is an association of entropy to the brain capacity for information processing, and it relates too to the speed of processing.

In a conscious brain the quantity of information processed is around ten times higher than during a resting-state. The brain large-scale information sharing is a consistent signature of conscious processing. Slow wave is associated with a reduced complexity at short-time-scale, as being generated by inhibitory activity.

In the last one hundred years the human speed of brain processing diminished by 10-15 percent. Since the Industrial Revolution, the speed of processing diminished by over 33 percent.

These estimates indicate that neural circuits capacity to transmit and process information has been reduced in a significant manner by an increased latency. Such **a decreased capacity in the speed of processing of information also indicates a lower entropy neural production**.

A lower entropy neural production shows a higher process of dissipation that brings more linear order to the brain functioning. Energy is dissipated when two surfaces rub together, as in flows subject to wall friction.

Friction increases the thermal energy of the circuits, and this energy becomes wasted, or it becomes ceded to the environment. The longer the neural circuits become, more signal friction occurs that slows down the transmission speed of the signal.

In the meantime, such a time-elongation processing decompresses the perceptual information that becomes more detailed. **Unveiling of these additional details is the "discoveries" we make. These "discoveries" built up our "linear" civilization**.

Most of the studies hypothesis that "latency" diminishes the "reaction-time". Other studies indicate that the deceasing of the "reaction-time does not always occur. Hence, the alleged mental retardation might not be so severe as initially thought.

In my view, the "latency increasing" is embedded in the process of "linearization" that tends to segment the perceptual reality into smaller and smaller parts, or into more detailed pieces of perceptual information.

The question here is about **what will be that limit of non-return?**

I have estimated that we are in a process of continual "linearization" that is the natural outcome of our modern brain anatomical changes. Symbolization became an innate to our modern brain because it currently manifests between the age of five month and before a child reaches one year old.

Is our future a slide into total individualism?

The aim of this paper is to investigate the evolution of individualism and how it may significantly influence our future.

The human brain evolution has gradually turned toward a perceptual information filtering that generated linear thinking.

High maneuverability of linear outcomes allowed the brain to segregate the informational content into separated entities. In this, the entity of the self would become selected further and gradually turned significantly enhanced. The self would become central to new linear thinking and would separate the individual from its embedded community.

However, the process that generated linearity was anatomically initiated in the Homo sapiens brain 70,000 years ago, while some complex results started to manifest only 40,000 to 23,000 years ago (like a dynamic and coherent language, significant changes in cognition, higher brain plasticity). Since then, changes in the human brain processing continued very slowly and were expressed on many distinct paths.

Whatever the anatomical changes in the human brain did, all of them delt to reshape the information provided by emotion. All species, or almost all of them, relay 100% on emotion processing to evaluate the surrounding reality. If there is any kind of thinking, it relays only on processing the emotions.

To some degree, most species process a part of their emotional information at the conscient level. It means, some consciousness exists in all species, and even within the plant kingdom.

All these linear increases, which windingly affected mental processing in humans over the last 70,000 years (but especially in the last 23,000 years), **found their summation only 200 years ago** within the advent of

Western Driven Industrial and Science Revolution that instated individualism as its main expression.

Science promoted individualism as an educational and institutional asset of Western Culture, while spreading it around the Globe within the colonization process.

Individualism stimulated and reshaped latently existing concepts of freedom, democracy, and human rights.

* * *

I like to quote some of Prof. Dr. Bruce Hood ideas expressed during an interview in 2012, and which refers to his article The Illusion of The Self-published on May 22, 2012, as an interview with Sam Harris.

"*Bruce Hood* is a celebrated neuroscientist who at the time of the interview *was the Director of the Bristol Cognitive Development Center at the University of Bristol. He has been a researcher fellow at Cambridge University and University College London, visiting scientist at MIT, and a faculty professor at Harvard. He has been awarded an Alfred Sloan Fellowship in neuroscience, the Young Investigator Award for the International Society of Infancy Researchers, the Robert Franz Memorial Award, and voted as a Fellow by the Association for Psychological Science.*"

"*Most of us have an experience of a self. I certainly have one. But the experience is an illusion-it does not exist independently of the person having the experience, and it is certainly not what it seems. Experience of self-illusion may have tangible, functional benefits in the way we think and act, but that does not mean that it exists as an entity.*"

"*There is conscious awareness of the present moment that we call "I", but there is also a self that reflects upon who we are in terms of our history, our current activities, and our future plans. I think that both the "I" and "me" are actually ever-changing narrative generated by our brains to provide a coherent framework to organize the output of all the factors that contribute to our thoughts and behaviors.*"

"*Many studies revealed that illusions generate brain activity as if they existed. They are not real, but the brain treats them as if they were.*"

"***The reason that the status of reality cannot be applied to the self is that it does not exist independently of our brain that is having the experience. It may appear to have a consistency of regularity and stability that makes it seem real, but those properties alone do not make it so.***"

"*We certainly have more choices today to do things that are not in accord with our biology, and it may be true that we should talk about **free will** in a meaningful way, but that **seems irrelevant to the central problem of positing an entity that can make choices independently of the multitude of factors that control the decision. We cannot choose the factors that ultimately influence what we do and think.***"

"***Free will is certainly a major component of self-illusion**, but it is not synonymous. Both are illusions.*"

If both are illusions, are democracy and freedom other mental illusions?

As the most typical example, "*democratic determination means that the **wishes of those who are in the minority on any issue are irrelevant to the outcome rather than providing any significant voice in a result they will be forced to accept. Since we all are in the minority on some issues, that is hardly an ideal to aim for.***"

* * *

At the bottom of natural hierarchies is the linear hierarchy, where linearization assets local stability and analyses the behavior near an equilibrium point of a nonlinear system. In doing so, linearization is a filtering process where certain information of too much variance must be eliminated.

The vector of variance and linearity goes from the main space to subspaces, showing the path of the hierarchic decreasing. In this process a significant amount of information is lost.

Within the linear hierarchy several phenomena occur. Among them, the **process of discretization tends to divide a continuous object into finite members of discrete elements. It seems that discretization is the phenomenon that may lead to "individualization," while dismantling most natural collectives**.

At quantum level, entanglement is one of the fundamental characteristics of matter. It defines a fundamental collectivity that makes exist the microworld.

Biological studies of the last decade show that **many quantic properties cross the boundaries, becoming expressed in the macroworld or the molecular world**. As a result of the mentioned crossing, the **molecular world acquirers' "collectives" as its fundamental future that promotes the development of all self-organizing complexities**. Everything in this world is one type of complexity or another.

* * *

The neuroscience found that reasoning is produced by the frontal cortex, and the emotions by the limbic cortex.

Self-control is the capacity to alter the tendency to produce a dominant response by impulse inhibition. This implies a similar filtering mechanism with one that, **selecting the information from nonlinear surrounding reality, produces a mental linear approximation of it**. Hence, self-control arises in the process associated with the generation of linear thinking.

Human consciousness before the advent of linearity of **thinking was based solely on the interpretation of emotions. Such emotions were nonlinear and beyond any brain understanding, but their projection into primitive minds generated intuitive approaches and contextual estimates. Primitive minds evolutionally adjusted to nonlinear hierarchies, which were circular, and displayed only qualities**.

Social organization 70,000 years ago reflected a corresponding **nonlinear organization and behavior**, **where the community used a type of parallel thinking approach** by distributing environmental data to all members. And then, it intervened in a reduction pattern, where all parallel processes were combined to reach an overall result or solution. This is a process similar to what the Science of Complexity describes as the production of an emergent.

Before 50,000 years ago, the human collectives were producing camp or cave accommodation, collective foraging, temporal synchronicity, collective choices, division of labor. It was a world dealing with the surrounding nonlinearity by the means of self-organization, which created the collective approach.

Those nonlinear social and group organizations may seem to produce chaotic behavior, but in fact, that **nonlinearity gave rise to orderly structures and social organizations. However, social interaction had produced effects or outcomes, which were not proportional to causes, and which make them impossible to trace or understand**.

Because of this, the **Paleolithic individuals were only simple observants of the surrounding effects and processes taking place around them**. Their interaction with the environment was minimal, and mostly related to feeding, primitive tool making, and accomplishing other simple tasks.

As the linearity spread within the human mind, abstraction, numerology, proportionality, and symmetry became pertinent. It occurred for the first time the search to turn qualities into quantities.

* * *

During the migration out of Africa, **around 55,000 years ago, the Homo sapiens' brain was still nonlinear, but it was transiting toward increased plasticity, cognition, and incipient linearity**. The language was primitive or static. Hence, the **main part of this transition occurred out of Africa, during the process that spread humans all around the globe**.

The gradual filtering transition toward linear thinking, which seemed to show a significant increase from 50,000 to 23,000 years ago, introduced the **linear hierarchy and its associated dynamics**. It also implied a

cultural switching that changed the mechanism of social interactions guided now by gradual adjustment to the increasing manifestation of novel linear behaviors, mostly expressed by the development of self.

Primitive societies delt with qualities only. The comparison was the king at those times. Numerology was limited to the basic ten digits.

50,000 years ago, there was a nonlinear world in the human brain; but that one sensed the surrounding reality only intuitively, contextually, qualitatively, emotionally. Survival was assured by collective interaction and cohesion, where everyone must help any other one. It was one step or two evolved from the rest of the monkeys.

However, some traces of **abstract thinking** evolved as a random feature around 70,000 years ago in South Africa. Much later or 50,000 to 30,000 years ago the cave paintings of Borneo and France/Spain showed a more matured abstract expression.

The introduction **of linearity allowed the brain to process analysis and composition**, which were not possible in the previous mental frame of the mind. **Rationality originates into analytical approach of thinking**. Thus, **linearity allows to break the wholes into parts and to recombine the parts in a different ways**. Such linearity gradually evolved from a previous nonlinearity where only emotional thinking existed.

Linearity radically transformed the previous sign-language and primitive vocalization into an increasingly coherent speech. 50,000 years ago, the language had only substantives and adjectives, and it was tagged as a static language. Syntax was poor and incapable of conceiving multiples, nor compositions. No time flow was sensed.

40,000 years ago, the language acquired verbs and adverbs, which made it dynamic and coherent. It occurred a significant progress in syntax and grammar. However, the complexity of the language evolved by a large increase in the number of words and in the way they have been connected or related.

Such improvements of the language are dated by archaeologists 30,000 years ago, when it produced a **language revolution**. Before this occurrence, primitive language was nonlinear, emotional, and multimodal, **using an array of nonverbal means to fulfil communication needs**.

* * *

The dynamic coherent language, generated 40,000-23,000 years ago, reflected a state of brain evolution. The language continued to evolve all through prehistory and history, reflecting changes in the brain functioning, or brain environmental adaptations, like those reflected by the haplogroups.

Linguistic studies connect the regional grammar and syntax of a language with a particular development of the brain. In this case, the geographic factor had been crucial in shaping a local language that reflects the local environmental conditions.

Without being racist, some studies reveal that the **Indo-European language bearers show an increased abstraction capacity, open mindedness, tendency toward polytheistic believes, inclination toward art, music, and science**.

The Indo-European family of languages was developed by the R1a haplogroup of people (called Aryans), who gradually migrated from Siberian environments to Central Asia, and further on to Pontic Plaines, Caucasus, and Anatolia. From there they continued to spread to Europe, the Middle East, and South Asia. In their journey the R1a people crossed a large variety of distinct environments, making them to continually adjust to each and every change.

The Corded Ware Culture genetic analysis shows that 3000 BCE haplogroup R1a and R1b existed simultaneously in this geographic region. East of it, the Yamna Culture had predominant R1a haplogroup. The R1a expanded overtime toward west, making R1b migrate further to the west. 2500 BCE the R1a haplogroups migration stopped at the Rhine River Valley, but retreated later to Oder Valley, which drawled a genetic border that continues to exist today.

In sum, continual adjusting to distinct landscape and conditions of R1a and R1b haplogroups produced a forced brain plasticity, significantly increased cognition. Therefore, it fundamentally influenced these people's mentality and their spoken language.

In the end, two distinct cultures have been produced: one in the west and one in the east of Europe.

However, **the split of R1a into R1b has an important significance**. R1a originated in the glacial era (around 24,000-22,000 years ago in the Yenisei Valley), even though it survived the Last Glacial Maximum. This origin shaped their initial mentality and the associated behaviors. The scarcity of glacial resources determined a tight collaboration with the group. Later adjustments were added to the initial package of collaborative behaviors.

The R1b appeared during the deglaciation, around 14,000 years ago. It enjoyed a favorable increase in resources, which **diminished the need for group collaboration but increased the competition between the groups' members**.

Warmer climate influenced the brain functioning, favoring higher dynamics, more energetic attitudes, assertive, and competitive behaviors, and approaches. Here, there is the genetic origin of individualism, and individual freedom.

Until 5,000 years ago, the individuals of the R1a and R1b haplogroups were tightly associated, traveling together. **Gradually, R1b evolved further on its own genetic path, becoming more distinct in its competitiveness and assertiveness**. Some sources indicate that the R1b gradually imposed themselves as the leaders of the mixed groups.

However**, around 3,000 BCE, the R1b groups started separate from R1a groups**, moving westward toward warmer zones of Mediterranean and Atlantic Europe to fit better the genetic background.

There is no doubt that **R1b separation of R1a was caused by their distinct mentalities and their different concepts on leadership and**

social coordination and achievement. Maybe it also appeared a distinct orientation in the type of belief.

The R1b separation allowed them to create their own cultural world, where assertiveness and competition became dominant, gradually breaking the collective behavior of the group. Their culture favored individualistic approaches.

I can conclude that R1a individuals remained inclined to employ collective and cooperative behaviors. Their competitiveness remained in a low realm because is not socially attractive in this type of culture.

However, the R1a individuals have a similar brain plasticity and cognitive capacity as the R1b bearers. Even then, their path of reasoning is different, producing an uncommon prowess in arts, music, and cultural achievements. Dialectic concepts are common among these people.

* * *

As another aspect of environmental adaptation, one can find that when people live for a long time in arid conditions, such people would experience a specific environmental adaptation that lead them to stubbornness, immobility, stagnation, resistance to change. These people tend toward monotheistic beliefs.

* * *

Linguistic differences would indicate a developing tendency toward individualism or collectivism.

Due to linearity, the initial languages have incrementally accumulated many words, becoming extremely complex. Later evolutionary increases of brain plasticity allowed their speakers to operate linguistic reductions and refinements. Grammar and syntax reflect the environmental adaptation that shaped those people's mentality. All adaptions are operated in the emotional background. **Changes in emotional content guide the thinking pattern**.

It is said that language is a tool or a technology in itself. In this sense, the language and its associated mentality generated distinct approaches toward technology making because **emotional content is adjusted by environmental adaptation.**

It can be said that the **locality of languages helped people's orientation toward analytical or collectivistic thinking**, and in general, **it separated the Western Culture from the Eastern Culture**. As it appears, the separation line between these cultures is the Indus Valley.

* * *

Today's moder brain still relays 95%, or even more, on emotional thinking that is also embedded into the language. Such emotional thinking is nonlinear at its fundament. There is the possibility that emotions are preserved into high entropic hierarchies, where they are stable because they are surrounded by higher level of entropy. Emotions represent evolutionary adaptations in all species.

Due to increased brain filtering capacity in the last 70,000 to 23,000 years, the **emotions have been subject to a similar filtering that linearized them**. By linearization, the emotional fundament was altered and multiplied; the linearized multiplication of the emotions servs better human immediate adaptation to environmental and external changes and influences. **Linearization saved Homo sapiens from the extinction experienced by Neanderthals and Denisovans because it allowed an immediate increase in adaptability**.

One may estimate that only 5% of the nonlinear emotions changed into linearity outputs. However, **once emotions linearized, they significantly multiplied, and their expression changed**.

This change in brain capacity of filtering seems to be the result of an **exceptional development of the precuneus** and Frontoparietal Synthesis ability in humans in the last 70,000 years.

However, **nonlinear emotions produce intuitive thoughts, while the linear ones partially participate in the generation of the self**. Everything indicates that such changes in our brain functioning are quite recent because they are very little expressed in aboriginal people who inhabit New Guinea and Amazonia and who turned isolated only 50,000 years ago or less.

However, **linear approach is also responsible for the modern evolution of the consciousness**, meaning a significant enlargement of it.

The advent of agriculture gradually introduced a means of competition among producers. **Maybe the first manifestation of linear thinking occurred when people started to evaluate things quantitatively**, and this may have occurred at the beginning of agriculture some 8,000 years ago or even earlier. Also, agricultural practice has partially stimulated random individualism among lonely producers.

The first known form of currency appeared around 2,500 BCE in Mesopotamia. The first known mints date 650 BCE and 600 BCE in Asia Minor.

Agriculture stimulated crafting of tools and other utensils. It introduced the craftsman as a new producer along with the creation of artifact and agricultural market.

However, the institutionalization of slavery 3,500 BCE in Sumer can be interpreted as a significant increase in individualistic approaches, where some individuals take advantage of others, enslaving them for a personal benefit.

The invention of **writing** (3,400 BCE in Mesopotamia) can also be interpreted as an approach toward individualism by means of recording a quantity for personal benefit and use.

It was paralleled by the **war making** that is another sign that traditional social collaboration became broken, while being replaced by an **aggressive type of competition**.

The first signs of analytical thinking, tagged as democracy and logics, are recorded as a random but elitist process some 2,500 years ago within Greek Antiquity.

Some theorists try to demonstrate that the early Christianity promoted an analytic thinking, too.

There is to mention that **the mentioned events were recorded in Western Culture**. In the **Eastern Culture**, around 500 BCE and in parallel with analytical evolution in Greek Antiquity, it occurred the generation of **collectively oriented philosophies** of Buddhism, Taoism, and Confucianism.

While in the Western Culture evolved slavery, in the Eastern Culture evolved an Early Type of Feudalism that was significantly less aggressive than slavery.

However, **the rise of the marketplace 500 years ago** initiated a significant stimulation of the self by introducing the monetary value of all assets, inflicting the personal assets, too. The **marketplace production was oriented toward personal needs, stimulating individual consumption of self-desired products**. The self gradually evolved as the own asset with certain quantitative value.

Nevertheless, the analytical thinking inflicted the braking of the social collectivity into individual selves or parts.

However, all mentioned evolutions were gradual, while showing a very slow evolutionary path. The generation of the marketplace represented an acceleration of the individualistic mentality. But **a radical change occurred only 200 years ago with the advent of the industrial revolution and the birth of science**.

There are reasons to believe that the introduction of modern science and its associated scientific education were crucial to enhancement of prefrontal cortex, contributing to other abilities in the social domain, while intermediating transactions required to achieve collectively required goals, referred to as **social intelligence**.

* * *

The disintegration of our social fabric, named social atomization, implies that the basic unit of society is being broken down into small parts. In medieval and early modern times, community was the basic unit of society, in part due to the lack of communication and transportation technologies that would allow a person to live securely and independently in their community.

One could say that **advances in technology increased the premises, which led to individualism**.

This technological transformation allowed the gradual introduction of multiple effects, like massive migration, multiculturalism, globalization.

Globalization makes the individual a sponge that soaks up the norms and beliefs of whichever locale they find themselves in. In cause no clear and permanent individual identity without which the individual cannot truly belong to any one community. Without community, individuals have no culture to provide shared customs and understanding that create common link and trust between people. **The ties of the community dissipate, and the individual becomes the basic unit of society**. Liberalism (that originated among the thinkers of the Early Church) has its unit of analysis expressed as an individual, but neither the community nor the collective are involved.

Capitalism benefits from social atomization because the more individual agents, the more consumers, and thus the more opportunity for profit.

* * *

We are a nonlinearly driven machine. Our nonlinear mind controls a nonlinear body.

However, the **linearity of thinking tends to segregate mind from body**. It makes our body run on its own controls, diluting the mind involvement and its role in the wholeness of the mechanism.

Such segregation may have deep health consequences, where the body would be allowed to make its own choices out of the mind desires. The body would become more influenced by the environment, and it would react independently to such changes.

We are gradually losing the mind control over our own body, which may turn, at some point, catastrophically.

* * *

The **illusion of the self** continues to increase its complexity, spreading around the globe. It gradually replaces local collectivistic approaches with increasingly individualistic behaviors. It imposes competition where collaboration dominates. It increases everywhere the aggressiveness.

The corollaries of the self would replace local traditions and behaviors. It introduces everywhere **the illusion of individual freedom, and democracy**, segregating the collectivities, switching their world views and behaviors.

The **qualities of life become quantified** as their turn to relate now to monetarization that defines the value of the individual. The mind becomes driven to produce monetary values. The **self becomes an asset** producer.

The individualistic minds, self-oriented, are focused on creating **products which would serv the individualistic needs**.

The market economy generates a vicious circle increasingly dominated by the consumption of the individualistic needs. It depletes resources, destroys the environment and the embedded ecosystems, tends toward a continual increase in production, but which production rapidly becomes only waste.

The remaining sustainability is rapidly degraded to catastrophic levels. The complexity we create exponentially increases, but mental solutions become increasingly distant from any resolution.

We slide rapidly on our mentally generated linear hierarchy toward nowhere. The illusion of progress dominates our mind, but there is no progress in nonlinear natural sense.

* * *

As much as individualism grows, it increasingly segregates us from the natural surroundings. We become an artificiality, where our ties with nature diminish at alarming levels. Could ever such an artificiality become self-sustainable and independent from the natural surroundings?

It would be an impossible dream.

Our society's driver is energy. We may soon achieve the reach of thermonuclear energy, which will allow us to significantly increase energy production and consumption independent of fossil resources. But here, there is a nonlinear component, where the linear increase in energy production affects the planetary energy balance.

Whatever we take from the planet diminishes the intake of the other natural agents inhabiting our planet because we are all bound to the universal laws of conservation of energy. **Our increased availability decreases the other possibilities**, driving them toward no living resources, causing an accelerated extinction path. Or we must circumscribe all other living things into our own society, feeding them as we feed ourselves. We must replace natural ecosystems with artificial ecosystems.

However, colonizing Mars, we escape from saving our Earth's ecosystems, because on Mars we will be alone.

Nevertheless, we must consider that acceleration of the individualistic segregation will continue to increase, and this tends to gradually replace much of remaining social collectivistic behaviors. We may become an immense number of isolated entities, where **individual competition for surviving will become more and more assertive or aggressive**. Violence will dominate such a future.

The means of individual communication will change, becoming more like a computer game. Even now, half of the planet spends significant time gaming. Cultures will change, too.

Who would assert individual rights? Democracy requires communication and collective participation. **Individualistic freedom is the opposite of democracy because it denies collectivity where democracy must be expressed**. Self-realization ignores others and is always done at the expense of others. Choice is good for the individual but bad for society.

* * *

Individualism tends to promote differentiation and inequality that is against democratic views of social equality. Inequality facilitates abuse and suppression.

A more self-interested individual has difficulties taking collective action.

Choice must be made to be good for yourself but also for society.

Countries with a high level of individualism have a higher ecological footprint and are less eager to act. Such countries' politics are a mirror image of their local individualism. In the meantime, these countries individualism was significantly involved in producing colonialism that was the instrument that led to their immense prosperity.

A mindset valuing collective goals would favor sustainability and would be centered on ecological action.

If deep ecological consciousness fails is because one does not fully identify with nature, it makes to disrespect others and their choices

* * *

From the physical point of view individual identity is subdued to the whole. The Western world lives a fatal separation where dominates the narcissistic **cult of individual proliferation**. Organisms do not exist outside of one another. One's separation denies entanglement and denies the existence of others.

The conflict between the individual and society arises where is a conflict between the individual's rights and the community's needs because the needs of the **majority will always be more important than the needs of the minority**. **Democracy is based on majority rule**.

Free will is an illusion because the choices are determined by things outside our control. Here, we control the linear, but everything surrounding us is nonlinear and out of our control.

Self exists only because there are others, who are different and make the social plurality. Such a diversity is essential for the ecological sense of existence.

Individuals with more individualist orientation are subject to take an adverse climate-friendly action than those with collectivist behaviors.

* * *

Minimal funding and investment in the third world countries generate a minimal development that makes such countries sensible to famine crises, leads to increased poverty, instability, local abuses, terrorism, and civil wars.

Because most of the mineral resources are in the third world countries they are subject to big corporations' competition and political maneuvering.

Such minimal development of the third world increases the economic differentiation between developed and underdevelopment countries.

In the meantime, the Western wars produced against them further destabilize these undeveloped environments.

All the above elements generate massive waves of migration toward the developed world. The migrants remain in the receiving countries for a long time without proper documentation, with limited rights and health assurance, dwelling in improper housing. Their ability to move around is also very limited. They become a minority with minimal rights.

The migrants become the cheapest labor force possible in the developed countries, increasing local profits exponentially.

I mentioned this migratory phenomenon because it negatively affects our concepts about freedom, democracy, equity, egality, and human rights. In the meantime, these facts make us wonder about the true motifs, which have led to such a massive migratory situation.

According to UNHCR at the end of 2022 108.4 million people were forcibly displaced, 62.5 million were internally displaced, and 35.3 million refugees. However, from 2000 to 2010 in the US came 14 million migrants. According to the World Migration Report of 2022, the number of migrants was 244 million in 2015, and 281 in 2020. Remittance of these migrants in 2020 was $702 billion. But even then, China is top to $83 billion and India second with $59 billion.

However, the developed countries profit in 2020 due to migration alone is estimated to several trillions USD (probably $3 to $5 trillion). The estimates for 2022 are around 300 million migrants.

Considering 300 million migrants, who work in the developed countries with highest technology available for their type of work, and under skillful supervision of local specialists, their annual economic input, like similar to a GDP, should be in the range of $10-$12 trillion.

For example, Africa GDP in 2022 was around $3 trillion, while the Latin America and Caribbean GDP in 2022 was $6.25 trillion. Together this two continents cumulate approximately 2 billion people and have a combined GDP of less than $10 trillion.

As another comparison, India has the same size of population as Africa and a similar GDP ($3.3 trillion in 2022). If one combines Africa and India GDPs ($3.0 plus $3.3 trillion = $6.3 trillion), the result is similar with the combined GDPs of Japan and South Korea ($5.1 trillion plus $1.6 trillion = $6.7 trillion). In this case, the GDP of 2 billion people equals the GDP of 200 million east Asian people.

However, the migrants come from countries where the collective behavior is dominant. In the receiving but developed countries, individualistic behavior is dominant, and this fact will significantly alter traditional behavior of the migrant population.

* * *

People in contemporary society are increasingly addressed as agentic individuals, making them responsible individual aspects of their own life, but also, responsible for overall sustainability and ecological attitude. Such an individualization helps one shape its own life, while tending to ignore relevant socio-economic processes, scope, and power relations. The environment has become a direct subject of capitalistic logic and political maneuvers.

The once optimistic vision of individualism forwards today in a hyper-individualism, where prioritizing the self is condemned as egoistic and harmful to the collective and ecosystem. Individualization is lost in the context because social contexts no longer structure life and because they remain at the margin of social structure that is an invisible structure pushing for individualization. The notion of social context is disappearing, and collectivity has slowly evaporated as a concept.

Recent research on Covid-19 pandemic has demonstrated the crucial role of the context, where individualized society failed to address significant social solutions and produced immense economic deteriorations.

New research suggests that increasing individualism is a global phenomenon. Some research focused on the US cases found narcissism and high divorce rate as an individualistic outcome, but similar results were found in other countries that were not Western or industrialized.

In India was a strong collectivistic approach, recent increases in diaspora, in foreign education of younger generations, but mostly due to significant capitalistic development in national economy, drives Indian society to a sky-rocketed increase in individualism.

The case of India may explain that countries, which were not show an increase in individualist value were among the lowest in socioeconomic development. Hence, socioeconomic development drives the rise in individualism.

China has been an exception, showing a significant a decrease in individualist values, while it experienced a high economic growth.

However, recent studies on China society show too, increases in individualization. For example, it was found that while south and north China are collectivistic, the individualization becomes significant in Yangtse Basin and northeast China.

Japan has experienced rapid economic growth and urbanization and adopted many social systems based on individualistic concepts. Along with such socioeconomic changes, Japanese culture became more individualistic. Social values become more individualistic. Local psychology becomes significantly affected by social, economic, and ecological factors.

Because the speed of socialization tremendously increased, its influence is huge, and cultures are changing dramatically.

I would like to cite and quote Judith Butler, a professor emeritus at University of California at Berkely with her article To Save the Earth, Dismantle Individuality, published in Time magazine on April 21, 2021.

"*As we dismantle the rigid forms of individuality in these interconnected times, we can imagine the smaller part that human world must play on this earth, whose regeneration we depend upon-and which, in turn, depends upon our smaller and more mindful role.*"

Another paper I would like to quote is The Privatization of Hope by Ronal Aronson published by the Boston Review on April 26, 2016.

"*Privatization of Hope refers to political, economic, and ideological projects of the past two generations, including the deliberate construction of the consumer economy and then the turn toward neo-liberalism. Under attack has been the kind of hope that is social, the motivation to make the world freer, more equal, more democratic, and more livable.*"

"*Not only does this privatization weaken collective capacity to solve collective problems, but it also deadens the very sense that the collectivity can or should exist, as the commons dissolves and social sources of problems become hidden. It undermines solidarity as the glue of social life. All foster a scenario of detachment, in which each of us is free to ignore our sense of belonging to a larger society.*"

"*Privatization of hope, then, is not simply a matter of focusing energy and attention on oneself and one's family. It is the withdrawal of personal expectations from the wider world, the rejection of even a possible democratic solidarity on behalf of a collective life encompassing and fit for all.*"

"*Taking control of one's life contributes to the process of de-socialization and de-solidarization. It also points toward the displacement of hope. Not only are collective capacities to solve collective problems weakened, but also the very sense that these problems are collective disappears. Displaced onto individuals, hope takes on the aspect of an addiction that can never be satisfied.*"

"*Paradoxically, the expended self-as-entrepreneur is an impoverished self, less and less concerned with the development of knowledge, interests, and capacities and more obsessed with maximizing competitiveness.*"

"*In letting markets decide our present and future, neoliberalism wholly abandons the project of individual and or collective mastery of existence. The neoliberal solution to problems is always more markets, more complete markets, more perfect markets, more financialization, new technology, new ways to monetize. Anything but collaborative and contestatory human decision making, control over the conditions of existence, planning for the future, anything but deliberate constructions of existence through democratic discussion, law, policy.* "(this is a quote made by the article's author from Michel Foucault).

"*Today what command our attention is not the radical falsity of the privatization of hope, which denies everyone's deep social being, but its debilitating consequences.* ***We are collectively losing the ability to coop with the most urgent problems. People who experience themselves as random, isolated individuals will never find the wherewithal to understand or agree upon, let alone master, the reality of climate change or any other major problem.***"

Dan M. Mrejeru

The Role of neural functional latency in the evolution of the modern linear human brain

Abstract

The aim of this paper is to analyze the possibility that a high process of natural linearization occurred accidentally in Homo sapiens brain. All brains to all species produce some linearization that introduces an orderly behavior necessary for surviving. However, the Homo sapiens brain developed a particular type of hierarchical linearization that arises from accidental anatomical modifications in their brain cortical organization and influenced the cortical connectivity and functioning.

Linearization created the self and other associated "illusions" as a process of discretization and quantification of the qualitative aspects of the perceived reality.

Linearization became a total approach affecting at least in part all brain cortical areas and their functioning, or at least those networks which serve the consciousness.

I tried to demonstrate in this and other papers of mine that many aspects we usually give them a distinct causality, in fact, they strictly followed the natural hierarchical linearization our modern brain is experiencing for the last 70,000 years.

The "pool effect" additionally intervened to share the neural network signals equally between non-linear and linear neural circuits.

All such linearization products created our civilization culturally and technologically.

One of the little discussed aspect is the **induced "latencies"** in all sensorial compartments of the human brain. **Our modern brain shows**

huge "latencies" compared with other primates and mammalians, and they continue to grow.

The latencies had and have a significant role in what we call "**discovery**," producing art, science, and technology.

The article debates "**digitalization**" and "**Artificial Intelligence**" as aspects of quantification, where the **"linearization" shows its expansive and highly progressive power**.

The paper debates the limits the "linearization" can reach and their implication for our social future.

Discussion

What is brain functional latency, and which are its effects.

Response latency is defined as the interval between the time a stimulus is given and the time a respondent gives a replay to that stimulus. A shorter latency means the brain is working faster.

Kosuke Itoh, Nako Konoike, Masafumi Nejime, Haruhiko Iwaoki, Hironaka Igarashi, Satoshi Hirata, and Katsuki Nakamura published in Scientific Reports 12, Article number:1103 (2022) an article with the title *Cerebral cortical processing time is elongated in human brain evolution.*

The authors explain: "*An increased number of neurons is presumed to underlie the enhancement of cognitive abilities in brain evolution. The evolution of human cognition is then expected to have accompanied a prolongation of net neural-processing time due to the accumulation of processing time of individual neurons over an expanded number of neurons. Here, we confirm this prediction and quantified the amount of prolongation in vivo, using noninvasive measurements of brain responses to sounds in unanesthetized human and nonhuman primates*."

"*A longer time window for auditory cortical processing is advantageous for analyzing time-varying acoustic stimuli, such as those important for speech perception. A novel hypothesis concerning human brain evolution emerges: the increase in cortical neuronal number widened the timescale of sensory cortical processing, the benefits of which outweighed the disadvantage of slow cognition and reaction.*"

"*An important assumption behind this hypothesis is that a larger brain contains a larger number of neurons, particularly cortical neurons, which augments the capacity for complex information processing. Consequently, the evolutionary scaled-up human brain contains a greater number of cerebral cortical neurons than the African elephant, which has a brain three times larger than that of an average human.*"

"*An increase in the total number of cortical neurons implies that the number of neurons that support a specific, cortical function, e.g., detecting a sound event or identifying a familiar face, also increases. Then, the total neural-processing time for executing a brain function is expected to become longer unless(1) all evolutionary acquired cortical neurons operate in parallel with previously existing neural circuits, or "A longer timescale of cortical processing allows neural inputs to the cortex to be integrated over an extended time window. unless (2) the temporal integration time of individual neurons (i.e., the time required by a neuron to summate synaptic inputs before it elicits an action potential) is shortened with the increase in the total neuronal number. To the best of our knowledge, there is no evidence to support these assumptions.*"

"*A longer timescale of cortical processing allows neural inputs to the cortex to be integrated over an extended time window. In the case of audition such elongation would enable analysis of the temporal features on time-varying acoustic stimuli that are important for the perception of speech and other complex sounds.*"

"*The evolutionary prolongation of cortical processing time has occurred not only in audition, but also in other sensory modalities, as the latencies of visual evoked potentials are also shorter in nonhuman primates than in humans.*"

As appears, higher brain functions are supported by longer timescale of processing, and higher tier cortical areas show extended timescales of neural processing in human brain. It may provide a clue to understand human intelligence.

However, the elongation of the timescale provides the details of the perceived reality, which becomes essential for surviving of larger mammals. **Such details are usually unseen without the occurring timescale prolongation**.

"*The peak mismatch negativity (MMN) (*for the auditive output*) latency is 150-250 ms in humans, 125-210 in chimpanzee, 80-110 ms in macaques, and 60-70 ms in marmosets*."

"*Studies have found that in children aged 4-6 there is less disturbance of speech than in children ages 7-9 under a delay of 200 ms. Younger children are maximally disrupted around 500 ms while older children around 400 ms. A 200 ms delay produces maximum disruption for adults. The delay required for maximum disruption decreases with age. However, it increases again for older adults, to 400 ms*." (Wikipedia)

The Plank Institute the language development branch indicates that between speakers the delay can increase to 1,500-3,000 ms.

Latency and linearization

In the case of humans, the brain anatomical changes drove brain functioning into a linear hierarchy, where it promotes linearization that assessed a local stability and analysis the behavior near an equilibrium point of nonlinear systems. In doing so, linearity removes, during functional inhibition, a significant amount of nonlinear content from the perceived information.

To compensate for this lost, **the same process of linearization operated a timescale elongation, which generated an additional amount of linear details**. It implied **a process of discretization, where it divides a continuum object into finite members of discrete elements**, and which become the additional details of the reality, and where **the variance turns limited, while divided by the discretization**.

The studies on analytical and holistic thinkers found that **analytic thinkers tend to focus on smaller details** available for their conscious processing, while the holistic thinkers do not. It implies that **analytic thinkers show a more linearized approach that involves larger latencies**.

However, in modern human brain terms, the analytical thinkers exploit preponderantly the left hemisphere, while the holistic ones are based mostly on the right hemisphere processing.

I can postulate that environmental causes, like frequent catastrophic flooding in south and east Asia, had prolonged collective action of communities, preserving a pre-existing holistic/collectivistic ancestry. This situation was not typical for Europe.

In this sense, I could say that **linearization is more advanced in analytic thinkers and in their left hemisphere**.

This is what we call **analytical thinking**. In the meantime, linearization favors **a new feature that allows to linearly combine the discrete elements**. We call such feature as **frontoparietal synthesis**.

Weakening of the sensorial input made humans develop the language.

Christer Johanson and Per Olav Folgero in their study *Is Reduced Visual Processing the Price of Language*?, published in *Brain Science* on June 12, 2022, indicated that "*the* ***linguistic abilities of visual thinkers may be impaired****, which suggests a negative correlation between visual perception/memory and language. What seems to happen is that* ***the reliance on primary sensation has weakened, which has ameliorated the development of high cognition***."

<u>The lowering of the intensity of neuronal information has been associated with shrinking of the brain volume</u>, and it coincided with a revolutionary development of language around 30,000 years ago, as a distinct modality of communication.

"***Human performance can be enhanced by altering the flow of information****, as in transcranial stimulation studies*." The transcranial technique (magnetic and electric) alters the neuronal connections and neuronal excitability, while it might lower the speed of the information flow.

Distinct linear approaches between Western and Eastern cultures.

The resulting brain redrawn-neural-connectivity caused by external stimuli favors the neural organization. It causes different mental ways for individuals to deal with the continuum of reality. Such distinct mental approaches generated distinct cultures. Thus, each culture shapes reality differently and significantly.

The Western culture individuals are associated with vertical and horizontal individualism and vertical collectivism. Easterners are associated with vertical and horizontal collectivism and horizontal individualism.

However, the analytical thinking of the Europeans allowed them conquest the world as a purported process of discovery, which turned instead into conquest and colonialization. During such colonial epoch they have built up for themselves an enormous economic-financial power. This power continues to impose today analytic thinking around the globe by the intermediation of the economic system of the marketplace.

Is latency responsible for discovering the unseen reality?

The gradual alteration of the information flow speed, causing latency, becomes essential in analyzing reality based on an elongated timescale modality.

Slowing the flow also divides it into discrete finite parts. Such discrete finite elements are normally those unseen details, because before the discretization, information is displayed as a continuum (or holistic).

The process of discretizing wholes (or continuum objects) was stimulated by the rise of the marketplace, where art and scientific knowledge became assets with significant social value. The **marketplace initiation was, in fact, an aspect of quantification of discretely drawn entities that quantified them with monetary values**. It was a major step toward further quantification of reality. It was followed by the quantification produced by science concepts based on mathematics, and it allowed a further quantification as industrial products with determined monetary quantitative values.

The Westerner discretizing, based on extending latencies, helped to discover science and art and drove toward Western rediscovery of the world. As we know, the impulse to discover the world and its mysteries dates from the Antiquity Epoch and continued in Medieval Asia. In fact, it was a circular development.

As one can observe, in the last 500 years, **it occurred in the Western culture a convergent path toward linearization, where quantification was the essential linear phenomenon**.

Because of colonialism and marketplace evolution, the quantification process spread around the world.

I might conclude that the Westerners are more oriented toward discovering more unseen details than the Easterners, while missing the relationship between them, but also they constantly missed the emergent behavior of every complexity that surrounds us. Easterners tend to evaluate complexities, like holistically, emergent behavior, and complex dynamic evolution (like contextually).

Several psychological studies point out the idea that there is a “general self” and a “contextual self.” We have to define which of these aspects is more important for our future doing.

Our world has all around natural and artificial complexities, which need to be addressed urgently to help us survive. **The discretization lacks perspective, and this should drive us to second thoughts about our future analytical policies**.

Digitalization and brain functional latencies

In the last two hundred years the Industrial Revolution was followed by **an increasing number of knowledge revolutions** (scientific and technological ones). All of them were **based on further discretization and quantification, showing an accelerated path toward linearization**.

It implies that our functional brain latencies increased significantly in the course of the last two hundred years.

The last come are the **digitalization and Artificial Intelligence revolutions**. The first one converts reality into numerical values, eliminating most of qualitative features of reality. It transforms the relationships developed within complexities into plain numerical interactions, eliminating most of the physical content, that is the fundamentally nonlinear aspect that governs them.

However, it revolutionizes communication in a numerical manner, allowing storing, analyzing, accessing, and sharing of data. It replaces analog interpretation with numerical one.

Linearization of mind illusions

The linear capacity to maneuver reality can create various **mental illusions, which cannot exist independent of the mind**, but which affect significantly human behaviors. We call them self, fee-will, freedom, democracy. In fact, they gradually become digitalized illusions, where the features of collectivity are broken down to individualistic illusional manifestations.

The will power manifest as a linear capacity to maneuver the linear entities. It acts by suppressing and/or enhancing or multiplying/reducing those syncretized elements we have created.

Linearization generated the mind illusions, where discretization created the oneness of self. The self, even illusory, attracted a lot of secondary, but illusionary too, aspects like self-free-will, self-freedom, and self-oriented democracy.

In all these illusionary aspects, linearization tried to quantify the mind, producing oneness to oppose that collectivity, which created such oneness (self).

To oneness have been attributed various illusory prerogatives, like freedom from collectivity, free-will that allow the oneness to perform independently of the collective, and democracy where oneness gets its independence from the faith of the collective.

All such illusory elements found a way to become some mental concepts with assessed social imposition. All they do is to quantify the mind in a natural way. It remains an illusory quantification.

As we can see, the path toward total linearity (as quantification) advances exponentially.

Here, I would like to make a brief review of human evolution.

Human evolution regressed into a linear hierarchy.

Due to significant increases in pulses of neurogenesis in the last 120,000 years, hominin brains were confronted with functional issues. The new neurons have been working mostly nonlinearly, and their functioning and connectivity was nonlinear, too.

The nonlinear neural working in the Paleolithic mind (that defined the primary consciousness) was intuitive, contextual, but hard to address in plain language. Increases in the number of neurons put the hominin brains in the situation that mental confusion (generated by proving at once multiple solutions to a common or complex problem) reached its apex and negated a clear but right solution for surviving.

Hence, the brain of Homo sapiens was forced to deal with this problem, and in contrast with other hominins, the sapiens developed the frontoparietal area with an exceptional evolution of precuneus.

As it seems, the **novel frontoparietal area in Homo sapiens has turned its functioning toward linearization**. Linearization made inter-neural activity and created relationships much easier.

It intervened the role of the "**pool effect**" that allowed one type of neuronal signaling to serve two distinct and opposing sources, like the old nonlinear neural circuits and new linear neural functions.

In my opinion **it occurred a hierarchical regression from basic nonlinear into basic linear functioning**. Such a phenomenon is not unusual in biology (like during their evolution some bacteria hierarchically degraded into viruses).

Thus, linearization of some of the neuronal functions saved Homo sapiens from the extinction process that affected the other hominins by altering their brain functioning and producing mental illnesses on large scale.

The saving of Homo sapiens was caused by the changes in the brain anatomy.

Hence, **today humans still experience a progressive conversion to linearity**, **which linearity was first widely manifested in the development of language (30,000 years ago), and the production of the self. Both contributed to moving the mental development from primary to secondary (modern) consciousness**.

Consciousness shows a totally linear functioning. Even when we have nonlinear thoughts (and we constantly have a lot of them), they are generated outside the consciousness realm and are not inserted in the consciousness functioning.

Since around 100,000 years ago, every step in human progress has constantly been a step toward more linearity. It manifested by gradual quantization of the previous qualitative properties manifested in behavior and functionality.

An obvious step backward of beneficial aspect of functional latencies (their prowess for discovery the unseen and unknown details of reality) is their produced slowdown of processing the perceive information. This increases proportionally with an increase in latency.

In sum, **as much as the number of details increases, our ability to process them proportionally diminishes**.

* * *

A lower entropy production shows a higher process of dissipation that brings more linear order to the brain functioning. As for flows subject to wall friction, energy is dissipated when two surfaces rub together.

Friction increases thermal energy of the circuits, and this energy becomes wasted, or it becomes ceded to the environment.

In the meantime, dissipation tends to discretize, a process that linearizes continual flows.

The neuroscience research indicates that resistance in white matter tracts is inversely proportional to the radius raised to the fourth power. It means that one tract constricts to half of its original radius, the resistance to flow will increase 16 times.

The brain-coordinated action depends on a dense network of fibers capable of rapidly transmitting information, both locally and to distant locations. The tracts of long-range fibers move information from these local functionally similar regions to areas separated by significant physical and functional distance.

The neocortex alone contains 220,000 km of short-range connections. The fiber length of the average cerebellar granule cells is around 330,000 km.

Faty insulating sheaths of myelin are used by the brain's longest fibers to increase conduction velocity at the cost of taking up more volume in the brain. There are 149,000 km of myelin fibers in the female brain and 176,000 km in males.

50,000 km of long-range fiber connects the cerebrum with itself.

Cerebral long-distance connections account for 20% of all fiber length.

White matter tracts are stretched by gray matter expansion.

There are many other mechanical factors in brain morphology. Highly organized strait trajectories of white matter tracts result from uniaxial tension that molded them in their final orientation.

The length of the tract is directly proportional to its resistance.. In the meantime, the more significant the opposition, the lower the flow.

The boundaries of the stretch-growth process generate tract elongation, which may be responsible for diminishing in tract diameter and transportation capacity that cause "latency."

* * *

Recent geophysicist 's opinions converge toward a more general description of the epoch 50 ka to 30 ka. It is divided into three phases:

-phase one is a stable phase dominated by axial dipole (50-43 ka) strength

-Laschamp excursion lasting cca. 5 ka (43 ka to 38 ka) and with surface field lasting cca. 2 ka (42 ka to 40 ka)

-weak phase, where the axial dipole and non-dipole power became comparable, leading to more than one excursion between 36 ka and 30 ka, but still called as the Mono Lake excursion centered at 34.5 ka.

As one can remark, the epoch 43 ka to 30 ka has been an era of significant geomagnetic activity, which corresponds with the most mental achievements recorded in modern brain evolution.

I mentioned two turning points:

-the transition from abstract to symbolic thinking (34 ka to 30 ka) demonstrated in cave painting

-the language revolution (demonstrated archaeologically) cca. 30 ka.

* * *

During geomagnetic excursions, the Galactic Cosmic Rays (GCRs) are charged particles coming from beyond our solar system that upon the interaction with the atmosphere produce a cascade of secondary particles among which the C14 isotopes and ultraviolet radiation (UV) are the most important ionizing elements.

The C14 Isotopes are absorbed by plants during photosynthesis, and such plants are eaten by herbivores, and herbivores are eaten by carnivores.

Many studies indicate a **significant effect of Low Doses of Ionizing Radiation (LDIR) on cells through oxidative stress**. Low doses of 0.2-0.5 Gy did not induce a detectable change in apoptosis compared to non-irradiated cells.

Mitochondria is a signaling hub in a cell, affecting the cells it is in and other cells through changes in mitochondrial proteome. Ionizing radiation at doses from 0.005 to 5.0 GY leads to an increase in the mass of mitochondria by 1.5 to 4.0 times. **In steam cells exposed to low doses ionizing radiation, metabolic activity increases, which may induce mitochondrial fusion that stimulates the stem cells proliferation and differentiation**.

However, **the ionizing radiation absorption produces ROS generation in cellular systems**, which, depending on the strength of the dose, can be good or bad.

Geomagnetic excursions have been an active element in our planet's history, and they never caused any known type of animal or biological extinction, which a high dose might have certainly generated. Thus, **excursions always have caused low doses of ionizing radiation that produced no harm to terrestrial life**.

A doubling of the concentration of the C14 isotopes in the atmosphere occurred during the atomic bomb experiments and it was identical in size with many prehistoric concentrations generated by the geomagnetic excursions and temporary pole reversals.

* * *

Kasturi Biswas, Kelianne Alexander, and Michael M. Francis, in their study *Reactive Oxygen Species: Angels and Demons in the Life of a Neuron,* published Neuroscience 2022, 3(1), 130-145 *(https://doi.org/10.3390/meurosci301001)* indicate that *"Reactive oxygen species (ROS) have emerged as regulators of key processes* ***supporting neuronal growth, function, and plasticity across lifespan****. At normal* *physiological levels, ROS performs important roles as secondary messengers in diverse molecular processes such as* ***regulating neuronal differentiation, polarization, synapses maturation, and neurotransmission****. In contrast, high level of ROS is toxic and can ultimately lead to cell death. Thus, maintaining tight control over ROS concentration in the nervous system is essential for proper neuronal development and function."*

Other studies enhance the essential role of ROS in **synaptic plasticity and memory**. ROS can play roles as secondary messengers, activating normal cellular processes.

Janel E. Le Belle, Nicolas M. Orozco, Andres A. Paucar, Jonathan P. Saxe, Jack Mottahedeh, April D. Pyle, Hong Wu, and Harley I. Kornblum in the study *Proliferative Neural Stem Cells Have High Endogenous ROS Levels that Regulate Self-Renewal and Neurogenesis in a P13K/Akt-Dependent Manner*, published Cell Stem Cell, 2011 Jan 7: 8(1); 59-71.

The authors indicate:

"*The majority of research on reactive oxygen species (ROS) has focused on their cellular toxicities. However, recent studies suggest that* ***ROS can also have proliferative, self-renewing multipotent neural progenitors with phenotypic characteristics of neural stem cells*** *(NSC); they maintained a high ROS status and were highly responsive to ROS stimulation.*"

"*The NOX-stimulated production of ROS, in turn,* ***can activate pathways that have been previously associated with enhanced cell proliferation and survival;*** *NOX isoforms have been identified in a number of different tissues, including or even especially in the brain.*"

"*In current manuscript we have demonstrated that both* ***exogenous and endogenous*** *ROS* ***can have a significant impact on neural stem and progenitor cell proliferation, self-renewal and neurogenesis****.*"

* * *

The Science of Complexity shows that every complexity generates an emergent behavior. I can assume that such a mysterious behavior, very distinct and apparently unrelated to that complexity's components, is just a hierarchical jump up and indicative of a distinct hierarchic physicality.

Hence, each complexity tends to upgrade hierarchically. The collapse of complexities is a downgrading aspect, where the components lose their relationships (or it is not enough energy to maintain them), becoming independent and isolated. After the collapse, self-organization will build-up a new complexity.

In a linear hierarchy is a hierarchical collapsing result encountered by a nonlinear hierarchy. It is different in scale from the resulting collapse of complex entities. Within linear hierarchy the interactions and relationships of neural networks become very simple because they are linearly maneuvered.

Even when our civilization appears as very complex, it is only a plain manifestation of progressive linearization. Of course, everything we have produced ends up as a complexity, but it was created by our linear manipulation.

The linearization path has its basic components as the increasing latencies, increased discretization, and behaves with progressive quantification.

Digitalization produced plenty of benefits in terms of efficiency. However, digitalization has to deal with a huge number of discrete elements for which it has to assure control by analyzing their stability and predictability. In doing so, digitalization falls prey to the procedural process where the procedures become lengthier and annihilate, in part, digital speeding up and closed looped gains.

However, **linearization is a finite process with a finite number of discrete elements. It maximizes until it reaches a totality of discrete elements, which are isolated and show no interaction. At this point the linear system of society collapses**.

Another study of a large team of researchers (led by Xiaodong Zhuang and Isabela Pedroza-Pacheco), published on July 14, 2020, in Communication Biology 3, Article number 376 (2020) indicates:

"*Our study supports a model where the low-oxygen environment of the lymph node may suppress HIV (hypoxia inducible factors) replication and* ***promote latency and viral development*** *and the* ***proliferation of HIV in a number of diseases, including cancer and inflammatory conditions***."

To conclude, functional brain "latencies" become associated with and favor functional body latencies, like hypoxia that promotes the spreading of various diseases, and cause body obesity.

Conclusion

Modern brain of humans originates into an accidental (at least accidental in its part) reorganization of the brain anatomy in the last 70,000 years.

The reorganization allowed the brain to widen its initial linear processing. Since then, every aspect of the brain expression and behavior became increasingly linearized.

Such natural linearization processes are causing progressive discretization, increasing latencies, and increasing quantification of all qualitative features of reality.

It is the right time to occur a wide discussion among scientists and philosophers about some gained possibilities (obviously provided by linearization to our mental benefit) to avoid those found effects inured in currently accelerated linearization.

Two different modalities to approach discovery. How our "progress" can run simultaneously backward and forward

Abstract

The aim of this paper is to document the role played by Delay-Discounting in Western vs. Eastern cultures, and their undisclosed implications and eventual contributions to our problem-solving ability.

Discussion

Bokyung Kim, Young Shin Sung, and Samuel M. McClure, in their study The Neural Basis of Cultural Differences in Delay Discounting, published in Philos Trans R. Soc. Lond B Biol Sci., 2012 Mar 5; 367 (1589); 650-656.

(doi: 10.1098/rstb.2011.0292) show:

"*Recent neuroimaging findings suggest that differences in discounting may arise from differential involvement of either brain rewards areas or regions in the prefrontal and parietal cortices associated with cognitive control.* We find that the ventral striatum is more greatly recruited in Americans relative to Koreans. **This suggest that a cultural difference in emotional responsivity underlines the observed behavioral effect**."

"*Aggregate cultural differences in intertemporal valuation are likely to produce conflicting preferences and opposing opinions on optimal courses of action across a broad span of situations. Preferences related to time affect nearly every aspect of our lives.* ***Everyday decisions commonly involve selecting between a more proximate outcome and another that will be realized only after an extended delay***."

"*One can hypothesize that Easterners are less sensitive to immediate reward than Westerners, because they have long-term perspectives that emphasize the future. Easterners do not behave as impulsively as Westerners even when faced with equally tempting immediate outcomes.*"

"*Using functional magnetic resonance imaging (fMRI), we found that there were no differences in brain responses in areas associated with executive processes. However, there were significant differences, suggesting that differences in intertemporal preferences derive from emotional responsivity to the presence of immediate rewards.*"

As is known, haplogroups reflect environmental adaptations. Hence, I can introduce a comparison about the behaviors created by adaptation of each haplogroup to environmental conditions.

The O haplogroup originated before the Ice Age Maximum cold (LGM) and evolved through such LGM frigid constrains. The same situation applies to haplogroup R1a. During the LGM the climate constrained the available resources for surviving, making the individuals adapt by enhancing collective and sharing approaches. Thus, the long-term perspective was another related adaptation.

However, the haplogroup R1b originated around 14 ka that corresponds to deglaciation epoch, where plenty of resources became available, encouraging competition and "immediate reward" approaches.

In the meantime, the environmental conditions of deglaciation have favored individualistic orientation toward resources; the impulsiveness, as a primary form of violence, has evolved further as a result of competition and conflictual situations implied by the race for optimal resources.

The haplogroups created before deglaciation ("O" and "R1a") have preserved their collective and contextual glacial-type-of-adaptations, which implied only minor alterations in behavior during changes into deglaciated environments; glacial era behavioral adaptation, as **collectivistic behavior, was easy to adapt with minor changes to new environmental situation.** The R1a haplogroup was almost born (cca. 27 ka appeared haplogroup R on the western shore of Baikal Lake, and cca.24 ka on Yenisei Valley, into one of its southern climate refuge near current Krasnoyarsk, it appeared haplogroup R1a) and matured within the LGM (24 ka to 19 ka).

I can say that **glaciated and deglaciated conditions generated two distinct cultures**. We still have them today as the avatars.

* * *

Analysis of R1a and R1b cohabitation reflects the birth of R1b around 14 ka in a still controversially defined area, which might be the Eastern and South Urals vs. Caucasian Region. Because of its later origin, the R1b-s remained for a long-time a minoritarian group of individuals cohabiting with the R1a in the places they migrated to.

Around 14 ka, the R1a haplogroup inhabited the North Pontic Plaines (current Donbass and South Ukraine), after they moved out of the Western Siberia climate refuge. This refuge was caused by the ice sheets damming the Obi River mouth at the Arctic shores. The damming created upstream an immense lake that warmed up the local climate. Naturally, the R1b-s became a part of R-1a-s, even though their genetics had been different.

The analysis indicates that, after a short while, the R-1b-s competing behavior manifested the tendency to impose themselves as leaders of the mixed population.

Even though R1a and R1b haplogroups cohabitated for a long time period of almost 10,000 years, the **R1b remains for the entire interval a rising minority that gradually took the overall leadership**. The R1b distinct behaviors accentuated in the course of time, eventually generating conflictual situations that led to their separation around 4 ka.

The R1b minority possessed a post glacial genetic outcome that favored a warmer climate. This explains constant Aryan migrations toward southern regions, like Central Asia, Grece, Middle East, South Asia, Egypt. At some point, around 4 ka, the R1b-s opted for a separation from R1a-s and moved to the warmer environments of Mediterranean and Atlantic Europe. The R1a-s opted to preserve their original habitat characterized by a continental climate with sharp while excessive seasonal variations.

This mentioned habitat matches largely the borders of the ex-Soviet Union from where it spread in Central and Eastern Europe the Slavic tribes that originated in the Volga River Basin.

I would say that R1a shows an intermediary genetic adaptation that can be placed between the behaviors of haplogroups "O" and "R1b". This is evident in South Asia where R1a and R1b migrated together, while **the majoritarian R1a produced in this region a "collective-type of culture" similar to that produced by haplogroup O in East Asia**.

* * *

However, in India, tendency of minoritarian haplogroup R1b to lead, while discretizing the continuum context of reality, surfaced in the **Hindus introduction of social castes**. Even then, the today Brahmins are R1a.

The creation of Buddhism in the epoch when the most of Nepal was part of India, it occurred just across the current Indian border. It indicates that majority of Hindu people were inclined toward "compassion," as their inherited collectivistic approach.

However, Buddhism had an orientation that contradicted the practice of castes. Hinduism officially accepted the Buddhism for its philosophical value, but the Hindu masses were discouraged to practice it.

There is a controversy about how the castes principles were applied. Recently, it was disclosed a speculation that Indian population cared less about the rule of castes, but the British colonial rulers favored the idea of the castes, while promoting, and even reinforced it for administrative and dominance scopes. It was easier for the colonial authority to deal with few influential leaders (especially the Brahmins and the few of upper castes) that to coop with large masses.

More collectivistic orientation among the East Asians made those individuals to adopt the Buddhism fully and rapidly. It first crossed into Tibet, and from their spread into China.

Even originating both in collectivist behaviors, Hindu syncretized between discretizing and promoting collectivism, while Eastern Asians tended to continually adapt their inherited collectivism to various novelties. From China, the Buddhism became rapidly embraced by Koreans, Japanese, and Southeast Asians.

The Islamic expansion and conquest into the east significantly affected South and Southeast Asia. Here, Islam took advantage of the misery status of the majoritarian population of lower castes. As a result of obtaining local political power, Islam inserted, in several areas where was dominant a Hindu spiritual orientation, its own Muslim competing belief that also promotes "an avatar collectivism."

* * *

One can draw the border between Western and Eastern cultures somewhere between Mesopotamia and Persia. The slavery originated in war-prisoners, and whose individuals in debt to creditors, who were forced work until the debt, or the guilt is paid off.

Slavery appeared in Egypt, Babylonia, and Media (but not in the rest of Persia) on the transition from 7th to 6th Centuries BCE. However, **slavery existed during Mycenaean Civilization since the 17th Century BCE**.

In general**, mass slavery as a whole has never been practiced by Persians. India and East Asia preserved for the entire period of ancient history a type of Feudal Order with no slavery practices**.

* * *

The slavery and war-making are both an indicator of violent behaviors and a **tendency toward individual suppression**. As Attica was promoting individual freedom, the Mycenaeans promoted individual suppression. Both behaviors became part of local cultures.

Suppression is mentally related to neural inhibition. Inhibition occurs when multiple alternatives, which populate nonlinearity and its uncertainty, are mentally reduced to a single one alternative.

The logic linearizes, discretizing the continuum of nonlinearity. It reduces it to a controllable discretized element. However, humans or Homo sapiens have evolved from a Paleolithic uncertainty toward linearization, discretization, and logic/analysis/synthesis. This evolution reflects the sense of urgency, where many alternatives, which populated the Paleolithic/Mesolithic contexts, were traded in for a single but "urgent" one that caused a short-term gain and eventually matched an immediate surviving need.

Since then, the Western culture continues to repeatedly prioritize short-term gains, while the Eastern culture continues to consider a context where the complexity of uncertain ongoing changes will produce an emergent result.

In the meantime, there are many examples where many operated short-term changes have deteriorated the long-term overall solution. Our environmental, climatic, and sustainability current situations are self-spoken.

Not to mention that short-term goals cannot be inflicted into long-term goals, unless the long-term is explicitly expressed for.

* * *

There is well known that "uncertainty" is the motor of "creativity." Somehow "uncertainty" is largely associated with risk-aversion and failure. In the meantime, uncertainty is a stimulation that promotes a mind exercise (called mindfulness) capable of obtaining a successful or optimal change. In fact, **there are several experimental proofs that many alternatives embedded in uncertainty are quite fair to approach and are risk-free.**

Decision making is when one decides what to do about the future.

The future does not exist, and it must be created. The future is uncertain, or one can say that it is the "uncertainty" that is unpredictable. Creativity lays within some uncertain solutions.

The logic is caused by slowing down the speed of neural signals. This latency in the signal's speed is caused by neural tracts diminishing in diameter as a result of tract elongation/squeezing process. The **increased tract's friction, followed by dissipation, produces slower-speeds, and unveils more embedded details, which cannot be distinguished or read at usual higher speeds**.

However, the logic originates from Frontoparietal Synthesis (FPS) that involves spatial mental maneuvering in precuneus. FPS is significantly related to the brain's functional latencies.

Context is significantly more distinct than a discretized reality because it allows the observer to identify competing cyclic and non-cyclic tendencies (changes), which simultaneously manifest in the environment. It gives the observer a raw idea what goals he/she may have to prioritize for future achievements and to become open to novel experience.

"*Creativity describes an open mind. In Zen culture an open mind is a beginner's mind.*" "***A beginner mind is a mind that is willing to see everything as if for the first time**. Zen practitioners describe the notion that learning requires an empty cup. When full of what you already know it is hard to acquire new knowledge. Almost all creative thinking techniques involve the abandonment of rational, logical thinking.*" (quotes from Creativity and Uncertainty, posted on July 18, 2020, by hbgelatt).

* * *

Returning to cultural differences in delay discounting, such **delay discounting is direct proportional to linearization for the Westerners, but inverse proportional in the approach of the Easterners**. It suggests **more nonlinear functions inherited and further developed in the Easterners' brains**.

A significant number of Japanese studies **had analyzed the "emptying of the mind process**", occurring during meditation and Zen practices. They converged to the idea that **during the "emptying of linear**

thoughts", the practitioner inhibits linear processing, while this stimulates in turn the nonlinear brain processing that is responsible for "intuition" and "creativity". Both are processed by the frontal lobe activity.

Reducing the mental search to frontal lobe, it partially diminishes the need for the use of the long neural tracks; it also inhibits the neural hub of precuneus; it diminishes or even inhibits the role of latency/linearity and of conscious (linear) thinking and its "self-related" approach.

The above-mentioned aspect **relates to "discovery approaches**." From the linear and quantitative viewpoints, **discretization/linearization in the Westerners mind produces a significant amount of "latency that diminishes the neural speed of processing that is the phenomenon that unveils that unseen where the "discoveries**" are embedded.

Discretization of a continuity of variables into bins for statistical analysis is known to produce some loss of information, and it affects statistical power of analysis or the generality of the analysis.

However, there is the assertion that continuous random variables, and continuous functions exist. **All purported variables are actually discrete, up to the finite level of accuracy of the computational representation, and here it depends if one wants to apply the universal principle where smaller bins can always be aggregated into larger bins.**

On the same principle, mathematicians transform/reduce nonlinear behavior present in the ambient (larger) space into the linear behavior of an embedded subspace, only if one single characteristic (that must be placed on a raw common boundary) is shared by both of them. Such a characteristic is term "covariation."

Hence, discretization can occur if an existing covariation is present. The same principle also applies when quality is transformed into quantity.

When the Eastern practitioner inhibits the linearity, he/she inhibits the logic, the latencies, and the outcome of the "linear discoveries."

Here, if the covariance principle pertains through the Universe, then, **individualistic approaches are false, because it implies an "individual co-dependency with the environment**."

I can conclude that the Easterners' minds are more focused on long-term changes and their eventual implications. Their discoveries are thus physically limited by their cultural trend.

However, the matter of "discoveries" in the Easterners is projected to dressed-up the future. <u>**Their "discoveries" are intended to match the "emergent but long-term outcomes" of the future**</u>, and it is actuated by applying a delayed discounting.

Jonathan Y. H. Sim, Jan W. Vasbinder of National University of Singapore published a study (*An Exploration of Complexity Science and Classical Chinese Thought: The Potential for Ancient Ideas to Enrich the Modern Study of Complex Systems*) published in Research Gate in January 2020.

The authors explain: "*The ancient Chinese did not have the concepts for understanding complex systems as we do today, but they were intimately familiar with complexity and had ways and means to understand and prescribe solutions for dealing with complex systems in everyday life, be it on the individual level of health, or with harmony on the social level.*"

This and several other studies enhance the significant connection between Ancient Chinese Philosophy and the Science of Complexity, at least in terms on **similarity of the "emergent Chinese philosophical conclusions", which for Science of Complexity are the "emergent solutions to the studied problem**."

Scientists think that the Science of Complexity evolution might provide, sometimes in the near future, more relevant answers to almost everything we intend to understand and to all severe problems we have created, but which are the ones we must fix to correct our damaged surviving ability.

* * *

Back to how these two different cultures produce their "**discoveries**."

What the Westerns do is **neglect the whole of complexity and its emergent behavior** because they proceed by **discretizing that wholes into composing parts**, and eventually analyze a limited interaction developing among a limited number of parts.

We from the Western culture take advantage of latencies or even stimulate their production because we can then see the "unseen."

But this "unseen" is an existing part of reality, and it is not any real "discovery." Here comes the saying: **no one never discovered anything because everything had already happened before we were able to see it**.

The mentioned bins of reality we discretize indicate that such bins are "subsets" with a common "covariance" as those larger bins where they are embedded. This implies that nature is a hierarchy of bins embedded one into another. So, binning into the small is the reverse of a natural process where the transfer of energy/information, and complexification always occurs from small into large. The universe itself continuously expends.

By contrast, **Easterners tend to wait until an emergent behavior arises from a certain contextual complexity they observe**.

However, the **Easterners approach is a step forward**, but it is still limited by the size of the bin where the emergence arises. And for that approach **they stimulate their "intuitive capabilities**."

Because the "intuition" provides a much more useful result that the "logic approach". It gives a valid solution to an observed "complexity."

But, as I said, "intuition is only one step forward and still very distant from the capability to analyze a multitude of complexities.

Many specialists proposed that the prehistoric use of hallucinogens helped our ancestors stimulate "intuition" and increase the entropy of the brain.

Neuroscience research **relates intelligent performances with higher levels of entropy**.

But here, as I previously discussed, the **latency produces dissipation of energy on neural tracts that not only diminishes the speed of the neural signal, but it also diminishes the speed of mental analysis**. In short, **latency kills our intelligent capability**.

I have to say that the **delay-discounting is also a form of latency but a forward one** because it follows the entropic arrow of time.

By contrast**, brain functional latencies** (typically enlarged in the Westerners brain) **are backward and negentropic regressions because they help discover only the past occurrences**. So, the Westerner "progress" does not drive us into the future, but only shows us everything that already happened, and which is the past.

Employing a scale-related meaning, the past is immense because it goes to the singularity or the Big Bang. In opposition to this, our discovery about the future is extremely tiny and has a limited temporal value. It may show only an Easterners contribution that is not immediate because it is affected by the delay-discounting.

Therefore, our type of "reverse progress" evolves exponentially, while the complex problems we are creating evolve exponentially, too, but without resolution because the solution is in the future.

We decomplexify reality, but at the same time, **the increasingly smaller bins of the reality we create self-evolve by self-organization, making them interact in novel ways, and this action produces future types of organization** (with more and more emergent behaviors and more or much more complex at every step).

As strange as it appears, we are running simultaneously two pathways:

-one is fed by our "discoveries" about the already consumed events

-the other one is led by self-organization that composes our "discoveries" into new organizations, generating "emergent

behaviors" we cannot control; this is the pathway that prospects into the future, and where the Easterners will have a small contribution, too.

These two paths feed one into another, and the process increases the overall complexities which surround us.

In the end, I would like to indicate which evolutions will produce future favorable changes.

One of the issue is the possibility to transfer quantum energy to the molecular level. Possible here that quantum computing, that would transfer quantum information to the molecular level, would indicate how we can transfer quantum energy, too.

Another issue is human feeding based on artificial lab-grown foods. It would save the current slaughtering of animals and would gradually eliminate the cultivation of the type of feeding plants we have developed, drastically reducing their areal extension.

A recent development in the manufacturing of the Tunnel Boring Machines (TBMs), would allow us to create extensive underground transportation that would gradually replace much of the ground one.

The mass and economically efficient seawater desalination will provide enough clean water for coastal and inland areas, where currently a water crisis manifests. It will protect local aquifers from depleting and assure a reliable outcome for environmental restauration of natural plant and animal species. Water hydrolysis future technologies will contribute to provide the hydrogen needed by new technologies.

The transition to lab-grown foods will also curb the current immense use of water in agriculture.

On the ground, it is the promise of hyperloop mass transportation.

The Hydrogen economy may allow us to eliminate the big manufacturing pollution in steel, cement, chemical manufacturing, transportation, and other highly pollutant industries.

Such developments must reduce our areal extension and footprint, which alter beyond recognition the planetary environments. Such gradual but crucial reduction of manmade environments and their footprints will provide room for restoring natural habitats.

However, right now, everything I mentioned above is only a small glimpse into such radically driven possibilities, while current deterioration of the planetary environment continues at an exponential rate. It is evident that the complex problems we have created and continue to be generated at exponential rates, would be extremely difficult to resolve.

In the meantime, we cannot be sure that our resolve would come in time to save the miseries we have created and to prevent our self-extinction.

A reason to catch emergent behavior: it shapes our future.

Here, the role of curiosity is significant

Abstract

Emergence is embedded in curiosity. What is curiosity and how it works? How curiosity maneuvers our lives? Curiosity-based is exploration. It is the most significant element that can predict the future. Curiosity among the generation Z is analyzed.

Curiosity shapes "introspection differently and negatively compared to curiosity that generate "intuition."

Asking WHY compared to asking WHAT. Curiosity here is emergent and produces the context that is nonlinear and holistic.

Discussion

Emergent properties belong to all disciplines. It is a property as appears as the complexity of the system increases and it does not on its own individual parts, but on their relationships to one another. It cannot be predicted by examining the system's individual parts, but **it still can be predicted by understanding the parts and their interactions that cause relationships**. In short, the **emergent results from the relationships (interactions) developed among the parts (as the lower-level components or subsystems)**.

The parts are arranged by self-organization into a structure that is complex; such structure determines the behaviors of the system. Much of current research studies individual parts (by syncretizing the whole) not the system as a whole, where the emergent properties are collective, being the properties of the whole system.

Emergent property is the fundamental property of every complex system. Such property can be weak or strong.

The weak emergent property is predictable and explainable and arises from the interaction of simple components. Such a weak property is entirely linear, and it deals with linear interactions among components.

The strong property is unpredictable, unexplainable by understanding only the interactions of individual components. Its causality is impossible to understand. It is derived from nonlinearity of the complex system and its interactions.

Realistic environments exhibit dynamic phenomena, which gradually increases the entropy. **By visiting novel states, one increases the entropy**. Novelty-seeking intrinsic motivation would produce an associated type of behavior. All animals, when encountering novel environments, perform complex exploratory behaviors. Such **exploration is largely motivated by reward-seeking behavior**.

However, **curiosity-based reward-seeking is a powerful exploration and learning mechanism. Exploration implies a continual shift of focus, where many behaviors emerge only to disappear by being constantly replaced by shifting of the search. Many of such behaviors are accumulated and reused as skills**.

Novelty seeking appears in combination with other goals, which are mostly guided by reward-seeking.

The interaction of curiosity drives the intrinsic motivation to learn as much as possible. The **curiosity drive** can be the progenitor of many social behaviors as well as the **generator of emergent behavior**.

It is said that curiosity attempts to maximize learning by selecting actions that give the largest amount of information.

Erik Hoel said: "There is a sense in which nothing in science makes sense without emergence."

Emergent behavior is a function of evolution itself, where systems allow them to **evolve through some adaptive and genetic programming, where learning is crucial**.

What creates the "emergent?" **Emergent** behavior means when one maximizes a few action/rules that **create large amount of possibilities**. But here, the mechanism one creates must be capable of having all components or simple systems interacting with one to another and not in isolation. For example, an emergent gameplay is fun because the player finds something that wasn't intended. The player is a tourist not an explorer. **Emergent behavior is somehow similar to the intelligence of the group**, where dumb individuals may exist, but they do not influence the intelligence produced by the group.

Ian Leslie (in his post Seven Varieties of Stupidity) said: "*Quite often in organizations that do stupid things, it's hard to pin the stupid decision on any one person in retrospect, and they may be no stupid individuals involved.*"

Matthias J. Gruber and Charan Ranqaanath published December 2019 in Trends in Cognitive Sciences (doi.org/10.1016/j.tics.2019.10.003) the study *How Curiosity Enhances Hippocampus-Dependent Memory: The Prediction, Appraisal, Curiosity, and Exploration (PACE) Framework.*

The authors explain: "*PACE proposes that curiosity is triggered by significant prediction errors that are appraised. This cycle enhances memory encoding through increased attention, exploration, and information seeking and enhances the consolidation of information acquired while in a curious state through dopaminergic neuromodulation of the hippocampus.*"

"*Accumulating evidence suggests that curiosity states are related to modulations in activity in the dopaminergic circuit and that these modulations impact memory encoding and consolidation for both targets of curiosity and incidental information encountered during curiosity states.*"

"*Researchers generally describe* ***state curiosity as a motivational state*** *that stimulates exploration and information seeking* ***to reduce***

uncertainty*. State curiosity resemble states that fall under the umbrella of reward motivation, in the sense that information that resolves uncertainty can be seen as having a value comparable with other rewards.*"

"*The effects of curiosity and memory can be understood as emerging from a cycle that involves Prediction errors, Appraisal, Curiosity, and Exploration. This framework proposes that curiosity is first triggered by significant prediction errors that are appraised as an indicator of information that could be valuable in the future. This cycle enhances memory encoding through increased attention, exploration, and information seeking. Prediction errors might not be sufficient to trigger curiosity or could even have the opposite effect and induce anxiety due to the uncertain state.*"

"*A PACE cycle will be completed once uncertainty is resolved and curiosity is satisfied by closure of an information gap, or it may elicit a further context or information-based prediction error. Such further prediction error will start a new PACE cycle which can further benefit memory and promote knowledge acquisition.*"

The **individual or agent is here rewarded for taking actions whose outcomes are not yet well predictable**. But the individual must provide a self-motivation, curiosity-based way for exploration. Here, it is very important to provide "*an application of curiosity learning towards the identification and retention of emerging intermediate behaviors.* ***Retaining emergent behavior from curious exploration equips the individual (agent) with a valuable skill set to learn new tasks more quickly***." (Oliver Growth, Markus Wulfmeier, Giulia Vezzani, Vibhavari Daasagi, Tim Hertweck, Roland Hafner, Nicolas Hees, and Martin Riedmiller, 17 September,2021 published in Google DeepMind).

Goren Gordon, Ehud Fonio, and Ehud Ahissar published a study in JNeurosci on September 17, 2014, 34 (38) 12646-12661 under the title *Emergent Exploration via Novelty Management*.

The authors were focused on defining the phases in which curiosity forms curiosity loops and retreat loops. The first loop "*maximizes novelty during outbound explorations ensuring the emergence of exploration motor primitive that maximized novelty during activation.*"

"*Our model assumes that* ***after novelty becomes higher than average, the retreat primitive is activated. When novelty is lower than average, the next exploration motor primitive is activated.***"

"*Our new quantitative findings suggest that curious animals do not attempt to maximize or minimize novelty, but rather maintain a constant flow of novelty by switching between behaviors that increase or reduce it.*"

"***Novelty-aversive behavior is similar to goal-directed policy, whose goal is to return to a known safe state****. This requires information that has already been learned about the environment and is to be contrasted with other fear-related behavior such as freezing, which does not.*"

"The basal ganglia are good candidates for the location for novelty management processing."

Now, I would like to relate curiosity to human evolution. As several studies, quoted bellow, indicate, **curiosity was the motor that made humans gradually build large societies, culture, and a rich social life**, which were developed well beyond other mammalians.

"*Highly curios people discriminate between intimate and small-talk conversations as they expect to bond better with partners during intimate conversations. Less curious people expect a lack of closeness regardless of the situation. These findings fit with appraisals at the core of curiosity-people feel curious when they believe there is a high potential novelty and possess the skill to cope with this novelty*" (Silvia, 2005). ***Highly curious people show greater interest in investigation new uncertain situations*** *because of strong novelty and strong coping appraisals*" (Silvia, 2008).

In my last several papers, recently uploaded on academia.edu, I enhanced **the role played by "latency" in unveiling the previously "unseen."** Such "unseen" became "seen" when a larger display of perceptual information became processed. It occurred when the slowing down of the neural signal transmission in elongated and squeezed neural tracs was produced due to neuronal multiplication in the Last Clade hominins and partially in the modern brain. **This becoming seen information I termed in my papers the "discovery**."

Now, in the context of the information presented in this article, **the "discoveries" are the fundament on which "curiosity" operates**. It produces a higher "attention" than usual and that may produce more "discoveries."

The "curiosity" would maneuver the "discoveries", seeking to find a common pattern among them. Such a pattern would generate "something" that would be a combinatory approach intermediated by the mentioned pattern.

This "something" is something new that did not exist before its combinatory while mental creation. It does not exist in nature either. It contains parts, and the interaction between these parts has an "emergent" result that we try to control by designing that "something" to avoid uncertainty and unwanted behaviors. It should have a low entropy, too.

However, all our creations, generating a lot of somethings, still originate and relate to curiosity.

Curiosity-based is a method for exploration. Curiosity learning is oriented toward identification and retention of emerging intermediate behaviors. Retaining the emergent behaviors from curious exploration equips the individual (agent) with a valuable skill set. Learning becomes much quicker.

Curiosity begets emotional engagement. It attempts to maximize learning by selecting actions that produce the largest amount of information. This behavior is reminiscent of infants' developmental stage of discovering their own body and reveling, while getting high reward in their own control of it.

The property of detecting patterns within sensory data is well-documented in central neurons placed along cortical paths in charge to analyze visual information. The **neurons detect specific relationships among the**

patterns of activity, and when stimulated by a curiosity-based motivation, they detect the emergent behavior, which result from such relationships.

In a survey, 75 percent of professional futurists identified "curiosity" as the number one trait of effective tomorrowing.

In a study of Sarah Davanzo, titled *Gen Z's Curiosity Fingerprint for Future,* published on August 10, 2023, in Rolling Stone, and Rolling Stone Culture Council, the author indicates:

"*While pursuing a master's in education, my professors complained that GEN Z is less curious than previous generations of students. However, my research found that their curiosity simply manifests in ways older generations don't appreciate.*"

"***Gen Z is the hacking generation-hacking education, software, finances, and products****. They are experts in finding the most efficient ways to do things, embodying the idea of "working smarter, not harder." Their curiosity is a "sensing" and "doing" kind. After all, GEN Z is the product of growing up in a volatile, uncertain, chaotic, and ambiguous world. They have developed the survival skills needed to solve the seemingly insurmountable problems impacting the future*."

"*When I probed deeper, these individuals had multiple interests. They are multifaceted and intersectional, yet incurious about things outside of their sphere of interests. Apparently, the mental bandwidth needed to juggle a portfolio of interests left little room for curiosity about the world at large*."

Regarding hacking for profit, I totally disagree with Sarah Davanzo. First, the hackers are a result of the evolution of the market economy, where market degradation drastically lowered the moral relationships, allowing and somehow promoting "fraud" as a way of life, but which severely affects, or even destroys, the lives of others. "Fraud" and "misinformation", "celebritification" are becoming the main expressions that govern our society. It is the outcome of relative "increasing distance" between the average blue collars from significant wealth. The creativity generated by curiosity will promote successful start-ups, not hacking.

Such hacking individuals are not Robin Hood-s, and they are not involved in resolving any of the planetary complex problems we have created. In the meantime, they generate additional complex problems which cannot be resolved. They do not help redistribute the wealth by their hacking. On the contrary, they intend to generate fraudulent-based wealth only for themselves.

I experienced significant hacking from so called self-publishing companies, and I have first-hand knowledge.

However, I do not contest that the **hacking behavior** is not curiosity-based. But it **occurs in an anti-social manner, denies the collective role of the individual for social benefit, and it represents the further evolution of the "individualistic" stance**.

Now, I would like to refer to the article by Tasha Eurich (an organizational psychologist) titled The right way to be introspective (yes, there's a wrong way) published by Ideas.Ted.com.

"*I had run a study looking at the relationship between self-reflection and outcomes like Happiness, stress, and job satisfaction. I was confident that people who spent time and energy examining themselves would have a clearer understanding of themselves and that this knowledge would have positive effects throughout their lives. But to my astonishment, our data told the exact opposite story.*"

"*Anthony M. Grant discovered that people who possess greater insight- which he defines as an intuitive understanding of us- enjoy stronger relationships, a clear sense of purpose and greater well-being, self-acceptance, and happiness. Similar studies have shown that people high in insight feel more in control of their lives, show more dramatic personal growth, enjoy better relationships, and feel calmer and more content. However, Grant and others have also come to realize there's no relationship between introspection and insight. This means that the act of thinking about ourselves isn't necessarily correlated with knowing ourselves. And, in few cases, they've even found the opposite: the more time the participants spend in introspection, the less self-knowledge they have. In other words, we can spend endless amounts of time in self-reflection but emerge with no more self-insight that when we started.*"

"*Why does this matter? After so many years of researching the subject of insight, I've come to believe that the qualities most critical for success in today world-including emotional intelligence, empathy, influence, persuasion, communication, and collaboration-all stem from self-awareness. If we are not self-aware, it's almost impossible to master skills that make us strong team players. Introspection was associated with poorer well-being. In truth, introspection can cloud our self-perception and unlash a host of unintended consequences.*

After all, what better way is there to increase our self-knowledge than to look inward?"

"*When we examine the cause of our thoughts, feelings, and behaviors-which we often do by asking ourselves WHY? Questions-we intend to search for the easiest and most plausible answer. Asking WHY can sometimes cause our brain to mislead us. Another reason that asking WHY is not always so beneficial is the negative impact it can have on our overall mental health.*"

"***So, if asking WHY isn't so helpful, what should we ask? Asking WHAT instead could keep us open to discovering new information about ourselves. WHAT not WHY. WHY questions can draw us to our limitations; WHAT questions help us see our potential; WHAT questions keep us CURIOS; WHAT questions help us create a better future***."

In my opinion asking **WHY is fundamentally related to analyzing and logics, and obviously, it is entirely "linear." By contrast, asking WHAT refers to a CONTEXT that is nonlinear**.

A context embeds a larger picture, where various elements interrelate one with another. **The context may show the emergent relationship resulting from the interaction between elements**.

Context suggests a powerful outcome on cognitive performance with the right hemisphere being in leading charge of the **perceived complex picture as a whole**. Context is holistic.

As long as curiosity-based stimulates emergent encounters, it becomes a tool set to explore this "emergent" and provides solutions to the complexities in which we are embedded. It also generates powerful skills for such a type of exploration.

In the meantime, the outcome of the reward would be exceptionally beneficial for all of us.

However, we should be aware that the emergent continuously evolve because the complex interactions incessantly evolve, too. Our linear mind does not allow us to be a step ahead of such emergent evolution.

About the Author

Dan has a MS from the University of Bucharest (Bucuresti), Romania, from the Geology and Geography Department (1975).

Dan worked most of his life as an Engineer Geologist, becoming specialized in Geotechnical Engineering. He immigrated to the US in June 1985. Since 1989, he has been the owner of a Geotechnical firm in Illinois, USA.

As a scientist he self-trained in the Science of Complex Systems (that is considered as the backbone of interdisciplinary knowledge), Geophysics, Neuroscience, Biology, and Genetics.

As an Independent Multidisciplinary Researcher, he became a member of academia.edu in 2016, where he published 93 research papers. On his site academia.edu he has over 10,000 views and 950 readers.

He also published research papers in the UK, France, and Austria resulting from his participation to international conferences held in those countries.

In the US, he became a member of CGScholar and The Interdisciplinary Social Sciences Research Network.

Recently, Dan became a member of NAIWE (National Association of Independent Writers and Editors, USA), The Authors Guild Member Community, USA, and The Society of Authors, the UK.

Dan self-published 10 books of which three of them have multiple editions (The Story of Homo loquens-4 editions, Seven Papers-3 editions, and The Making, the Rise, and the Future of the Speakingman-5 editions). Most of his book are in the category of general knowledge.

Three of his books were literary attempts with philosophical debates and interpretations (The Limits, Seven Essays of Creation, and The Creativity Seminar). He is currently working on the second edition of The Limits, where he made a fictional story that somehow describes his own nonlinear personality (this book was his first book, originating in 1994, but published only in 2006).

His book Solovki Ersatz (2015) was his first tentative research into the evolution of the modern brain.

On February 1, 2024, his book The Making, the Rise, and the Future of the Speakingman-Fourth Edition (published on December 8, 2023) won a Gold Book Award with prestigious Literary Titan (Irvine, CA).

On May 26, 2024, his book The Making, the Rise, and the Future of the Speakingman-Third Edition (published July 25, 2023) won a Silver Book Award with the international prestigious Nautilus Book Award Co.

His website is danmirceamrejeru.com.

His email address is d.mrejeru@gmail.com.

Printed by Libri Plureos GmbH in Hamburg, Germany